MATHEMATICAL TOOLS *for* CHANGING SPATIAL SCALES *in the* ANALYSIS *of* PHYSICAL SYSTEMS

MATHEMATICAL TOOLS *for* CHANGING SPATIAL SCALES *in the* ANALYSIS *of* PHYSICAL SYSTEMS

William G. Gray
Anton Leijnse
Randall L. Kolar
Cheryl A. Blain

CRC Press
Boca Raton   Ann Arbor   London   Tokyo

Library of Congress Cataloging-in-Publication Data

Catalog record is available from the Library of Congress.

Direct all inquiries to CRC Press, Inc., 2000 Corporate Blvd., N.W., Boca Raton, Florida, 33431.

International Standard Book Number 0-8493-8934-8

Printed in the United States 1 2 3 4 5 6 7 8 9 0

Printed on acid-free paper

TABLE OF CONTENTS

PREFACE

This book has its origin in a graduate course on averaging theory for porous media flow taught by the first author at the University of Notre Dame. During one semester when this course was being offered, the students became fascinated with the manipulations that went into the derivation of appropriate averaging theorems, and the course detoured into a study of the use of generalized functions as tools for the development of a wider range of theorems useful in engineering analysis. The most commonly encountered of these theorems are the transport and divergence theorems used in fluid mechanics, as well as other fields, in the derivation and application of conservation principles. The material in this book goes well beyond these theorems and reflects the authors' belief that these standard theorems, as well as many others that find their utility in the study of the conservation laws, can most easily and effectively be developed in the context of generalized functions.

The book might best be considered to be composed of four parts:

- Spatial, surficial, and curvilineal gradient, divergence, curl, and partial time derivatives for functions are developed using vector notation. The presentation avoids the use of metric tensors and provides relations that are very useful in the development of integral theorems (Chapters 1 and 2).
- Generalized functions are presented and their use in identifying portions of curves, surfaces, and volumes is demonstrated. Manipulations with these functions are also shown to facilitate the interchange among integrals of a function and its derivatives over curves, surfaces, and volumes (Chapters 3 and 4).
- A few typical derivations of integration and averaging theorems are presented in order to demonstrate the techniques employed. Following this, fifty-six gradient, divergence, curl, and transport integration theorems are tabulated. Finally, seventy-two averaging

theorems (eighteen each for the divergence, gradient, curl, and time derivative operators) are presented. These 128 theorems provide a very powerful array of mathematical tools for rigorous derivation of conservation laws at various scales and for interchange among those scales (Chapters 5 through 8).

- Application of the theorems to point spatial conservation principles is demonstrated for some common cases. The vertically averaged mass and momentum shallow water equations used in tidal modeling are obtained. Equations for river flow (integration over a cross section) are developed that include terms accounting for the channel curvature. Conservation equations for an aquifer are obtained from the averaging theorem. The use of the theorems to obtain point surface balance equations in a general surface is demonstrated (Chapter 9).

In all instances, the approach is an engineering one with some of the formalities and notational shorthands of precise mathematics sacrificed in favor of accessibility. Additionally, an attempt has been made to make the book modular such that, for example, one need not work through the manipulations involving generalized functions in the early chapters in order to make use of the integration theorems collected in Chapter 7.

The approach presented here is not a traditional one. Those disinclined to add generalized functions to their arsenal of available mathematical tools will find little use for the early chapters of the book. Nevertheless, the catalytic nature of generalized functions, wherein they enter a problem to assist in making a mathematical transformation and then disappear from the final result, means that the resulting theorems collected in Chapters 7 and 8 will be pleasing, even to the traditionalist. The material in Chapter 9, wherein a few of the theorems are applied to change problem scales, is intended to provide only a glimmer of the wealth of applications for which the theorems can productively provide a change in scale. The authors believe that if one spends some time with the material in this book, an increased appreciation for the ease with which problems may be formulated in space, on surfaces, or along curves will result. We have found it an intriguing challenge to prepare a book in which we can share some useful information as well as our enthusiasm for the techniques developed.

We are grateful to Michael A. Celia who has provided us with encouragement and valuable feedback on various aspects of the manuscript; to Michael L. Freeman who prepared the technical drawings from vague sketches and conceptualizations; and to the many other colleagues and researchers who have personally or through their publications provided insights into various aspects of the topics under study. We appreciate the confidence that CRC Press, Inc., and its publisher, Joel Claypool, have shown in us by taking on this project and the production assistance provided by Rosi Larrondo. Finally, we are pleased to acknowledge NSF grant INT-8822198 and the encouragement and support of the Dutch Institute of Public Health and Environmental Protection (RIVM) that contributed to our ability to undertake this international project.

William G. Gray
Anton Leijnse
Randall L. Kolar
Cheryl A. Blain

Notre Dame, Indiana
March, 1993

CHAPTER ONE

SCOPE AND BASIC CONCEPTS

1.0 INTRODUCTION

This book is concerned with three useful and powerful tools for mathematical analysis of engineering problems: vectors and vector calculus, generalized functions, and integral theorems. While these topics have been incorporated into the study of the various engineering disciplines for many years, the results and the methods used to obtain results in this text are novel. This book is not another presentation of traditional applied mathematics methods or of mathematics for engineers. Generations of very good texts in this area exist already [e.g. *Dull*, 1926; *Wylie*, 1966; *Hildebrand*, 1976; *Greenberg*, 1978; *Kreyszig*, 1979; *O'Neil*, 1987; *Zill and Cullen*, 1992]. Neither is it a variation on the approaches to continuum or fluid mechanics followed in many fine texts [e.g. *Aris*, 1962; *Malvern*, 1969; *Bowen*, 1989; *Truesdell*, 1991]. Rather, the material herein introduces tools that are either extensions of existing techniques or that have not been previously applied in the context of engineering analysis.

Those seeking a traditional exposition or traditional application of mathematical tools in an engineering context will have to look elsewhere. The scope of material in this text is different, and the paradigm of theorem development does not appear in other texts. Since the material included represents a break from tradition, learning and understanding require both an investment of time and a willingness to consider new approaches. However, fruitful dividends await. The tools streamline analyses and proofs that were once, at best, extremely tedious and cumbersome, if not intractable. The techniques lead to useful theorems for change of scale as they also provide a new approach to solving problems. Because many of the concepts in this book are a significant break from traditional approaches, outlining the material in an introductory chapter relying on a reader's prior experience is difficult. Instead, this first chapter provides some perspective to indicate the need for the material and the context in which the tools and theorems will prove useful.

1.1 SCALE AND THE DERIVATION OF BALANCE LAWS

1.1.1 Microscale, Macroscale, and Megascale

This book is motivated by the authors' interest in obtaining appropriate conservation laws for environmental systems, specifically, continuum representations of mass, momentum, energy, and entropy balances. These laws are the basic principles underlying physics-based modeling of flow and transport. In modeling these processes, at least three spatial scales are of importance in describing the various dynamic mechanisms. These are referred to as the microscale, macroscale, and megascale. Briefly, the microscale refers to the smallest scale at which a system can be viewed as a continuum. Below this scale, the materials under consideration are viewed as a discrete collection of molecules or particles such that modeling approaches that do not explicitly involve spatial gradients would be required. Such molecular scale techniques go by various names, e.g., cellular automata, discrete element modeling, or Stokesian dynamics. *Frisch et al.* [1987] and *Brady and Bossis* [1988] provide general reviews of these methods. The next scale above the microscale, the macroscale, is larger than the microscale but smaller than the scale of the system under study. It often represents a scale at which a multiphase system can be idealized as set of overlapping continua. Finally, the length scale of the megascale is on the order of the system dimensions. If a system is modeled as being completely megascopic, spatial variations are not explicitly considered as the system is characterized only in terms of average values. Although the different scales can be conceptually described, the transition from one scale to another is not clearly defined and the scales cannot be identified with precision. Thus the scale of observation is often problem dependent with balance laws for a particular problem being required and implemented at the appropriate scale.

1.1.2 Approaches to the Derivation of Balance Equations

Derivation of conservation laws at any scale generally follows one of two approaches. The first, referred to here as the control volume approach, begins by defining an element of volume for the system under study. The size and shape of the element are selected to be convenient for the system. (When the control volume is taken to be stationary and of constant size, the derivation is referred to as an Eulerian approach. However, if the volume moves such that it remains fixed to a particular portion of the system, the derivation is said to follow a Lagrangian approach.) For example, if the shallow water equations describing the balance of mass and momentum in an estuary are being derived, then a convenient control volume is a rectangular prism spanning the depth of the fluid. Such an element would have its vertical faces aligned with a cartesian coordinate system. Once this element is defined, fluxes of quantities across the faces of the element are balanced with internal changes. The conservation law that results is an algebraic expression that may be reduced to a differential form by taking the limit as the cross-sectional area of the prism goes to zero. Note that for this case of shallow water flow, the height of the prism spans the full depth of flow such that variation in the vertical is not accounted for. Thus the scale in the vertical is megascopic, while the horizontal scales are microscopic.

Although the control volume approach does provide useful balance equations, it suffers from several shortcomings. The derivation of balance laws from first principles can be tedious and must be repeated for each system and each scale, or mix of scales, under study. Important terms are easy to overlook, especially when the element geometry is complex, non-cartesian coordinates are of interest, or different scales are employed in different coordinate directions. For these reasons, an approach that systematically allows for transformation between scales is desirable.

The second approach to derivation of conservation laws, the one adopted in this work, is the integral approach. Its formalism circumvents the shortcomings of the control volume approach by using mathematical theorems, referred to collectively as integral theorems, to affect a change in scale of a balance law obtained at a different scale. Prototypical classical examples of this set of theorems are the Gauss divergence theorem, the Reynolds transport theorem, and Leibnitz' rule. To develop balance laws at the microscale, the theorems may be used in conjunction with elementary physics and applied to material volumes. (Applications in Chapter 9 illustrate the procedure.) To develop macroscale/megascale balance laws, that is to affect a change of scale of the problem, theorems are applied to microscale balance laws - a process frequently referred to as averaging. The process of selection of an appropriate control volume for derivation of a balance law is replaced by identification of the length scales at which the system is to be studied. With the integral approach, one proceeds systematically to desired balance laws by a rigorous and unambiguous change in scale. Derivations are streamlined in that most of the effort is expended *a priori* in developing the theorems. Use of the integral theorems, collected in Chapters 7 and 8, precludes inadvertent omission of terms accounting for various processes.

1.1.3 Evolution of Integral Theorems

Use of integral theorems to derive conservation laws has been restricted in the past because, aside from the prototypical examples mentioned above, only a few additional theorems have been developed over the years [e.g. surface transport theorems as found in *Aris*, 1962; *Stone*, 1990]. Starting in the 1960's, this state of affairs began to change when volumetric averaging was developed as a tool for derivation of continuum equations for multiphase system. The goal of averaging was the derivation of the equations governing flow at the macroscale such that processes occurring at the interfaces were accounted for as bulk processes rather than through application of boundary conditions at those interfaces. Implementation of averaging required that theorems be developed that related averages of derivatives at the microscale to macroscale derivatives of averages. In particular, a spatial averaging theorem (a three-dimensional analogue to Leibnitz' rule) and a less restrictive transport theorem were needed. These theorems were developed independently by *Anderson and Jackson* [1967] in their study of fluidized beds, by *Slattery* [1967] in his study of viscoelastic flow in porous media, and by *Whitaker* [1967] in his derivation of the dispersion equation for mass transport in a porous medium. Other treatments of the spatial averaging theorem followed [e.g. *Drew*, 1971; *Bachmat*, 1972]. In these papers, different methods to derive the theorems are presented, but the underlying theme for each is a classical approach based on geometric arguments similar to those found in derivations of the Gauss divergence theorem and the general and Reynold's transport theorems. For example, the spatial divergence theorem is traditionally derived by identifying a volume, integrating each term of the divergence of a vector over that volume, and then converting the volume integral to a surface integral using projections of differential areas. Extension of this approach to the more complicated geometries encountered in multiphase systems is complex and has been inhibited by the arduous task of keeping track of deforming regions and interfaces between phases.

A decidedly different approach to the derivation of integral theorems was first presented by *Gray and Lee* [1977]. Their work provides the basis for a method that moves time or space dependence of an integration region from the limits of the integral into the integrand so that tracking of moving and deforming subregions is greatly simplified. Generalized functions are at the core of the technique. Their use allows theorems which have heretofore been untenable (due to the need for tracking moving boundaries in complicated multiphase systems replete with interfaces and contact lines) to be developed. Also, these functions provide a convenient means to interchange among integrals over curves, surfaces, and volumes. Over the years, the method has been refined and extended [e.g. *Hassanizadeh and Gray*, 1979; *Gray and Hassanizadeh*, 1989; *Hassanizadeh and Gray*, 1990] so that theorems for virtually any system geometry, scale, or dimensionality can be obtained.

Although an approach based on generalized functions is an extremely powerful tool for use in engineering, resistance to implementation of this approach remains high because of lack of familiarity with the use of such functions (also called distributions). Additionally, much of the material related to the usage of these functions in engineering analyses is scattered throughout the literature or has heretofore not been published. An explanation of the power of generalized functions in the derivation of theorems for changing scale that also provides enough insight to the properties of generalized functions for use in other applications is warranted. Furthermore, the value of collecting the theorems for changing scale in a single work is apparent so that these tools can be readily accessed as needed, perhaps in the same manner as the relations in tables of definite and indefinite integrals so common in mathematics handbooks are accessed. The present work addresses both of these issues by providing a set of 128 theorems and the techniques used in their derivation.

1.2 COMPONENTS OF THE BOOK

This is a book on integral theorems. As stated, the theorems can be used when a change from a point representation to a representation at some larger scale is sought, or they can be used to derive microscale balance laws from first principles. As with all theorems in mathematics, the tools used to prove each theorem and the logic behind the proof itself hold significant pedagogical value, but complete mastery of each proof is not a prerequisite for proper use of the theorems. Instead, the theorems provide simple tools to change the scale of a problem involving almost any geometry in one, two, or three dimensions. While recognition of the value of the theorems in this work had its origin in the derivation of continuum laws in physics, the use of these theorems is not restricted to that discipline. Rather, they can be applied to any problem satisfying the hypotheses used in their derivation. The applications in Chapter 9 provide a sampling of the diverse use of the theorems.

This is also a book on the use of generalized functions in the context of theorem development. In short, a systematic framework may be defined in which generalized functions are used to derive integral theorems. In a very real sense, generalized functions are mathematical catalysts - they facilitate the derivations but they do not appear in the end product. As in the context of chemistry wherein a reaction would not take place without the presence of a catalyst, the derivations here rely on the catalytic activity of the generalized functions. The advantage of this approach is that theorems for complicated geometries, such as contact lines in a multiphase fluid, become tractable. In fact, the mechanics of the proofs of each theorem are the same regardless of the complexity of the geometry or the dimensionality of the derivative operator, although the derivations become lengthier with system complexity. The exposition on generalized functions extends the familiar one-dimensional Dirac delta function and Heaviside step function, neither of which is a "function" in the strict sense of the word, to their multidimensional analogs. Additionally, the derivative and integral properties of these quantities are provided with emphasis on their power to interconvert integrals over curves, surfaces, and volumes.

This book uses a consistent vector operator notation representation to indicate spatial, surficial, or curvilineal derivatives. In order to work with arbitrary system geometries, a need arose for a convenient means to represent vectors and the calculus of vectors without resorting to the conceptually difficult approaches of classical differential geometry including covariant and contravariant differentiation. The adopted vector representation and its calculus is a useful and powerful tool in its own right.

1.3 ORGANIZATION OF THE BOOK

The nine chapters of this book cover four primary subjects: vectors and vector calculus, generalized functions, integral theorems, and applications. Each chapter begins with an introduction that discusses the objectives of the chapter, continues with a few sections that accomplish these objectives, and closes with a summary of pertinent results and list of references.

Chapter 2 is devoted to mathematical preliminaries, primarily vectors and vector calculus. As mentioned, system geometries may not necessarily be cylindrical, spherical, etc. so an arbitrary orthogonal coordinate system is introduced; and corresponding differential operators are defined. Additional identities and relations among the operators are also established. Chapter 3 defines generalized functions in one, two, and three dimensions and establishes their integral and derivative properties. In Chapter 4, scales at which physical systems may be analyzed are defined. Concepts such as continuum representation and representative elementary volumes are discussed. Also, Chapter 4 identifies the two broad classes into which the integral theorems are grouped: integration theorems and averaging theorems. In very general terms, averaging theorems are used to affect a change of a problem from some small scale to a larger one while accounting for variation of properties in space at the two scales. Thus averaging transforms a continuum representation at a small scale to a continuum representation at a larger scale where the quantities subjected to averaging are essentially filtered or smoothed. Integration theorems, on the other hand, affect a change from the microscale to the megascale such that a continuum representation is transformed to a lumped representation. Of particular importance in Chapter 4 are guidelines on the use of generalized functions to transfer space and time dependence from the limits of integration to the integrand and use of generalized functions to change the dimensionality of the integration region. Concepts are illustrated with examples. In Chapter 5, a compact and utilitarian theorem notation is presented to facilitate theorem identification. Chapter 6 establishes useful identities involving generalized functions and then uses the suite of tools developed to detail the derivation of several representative theorems. Examples are chosen which illustrate subtle use of the tools. Chapters 7 and 8 contain listings of the theorems with Chapter 7 presenting integration theorems and Chapter 8 presenting averaging theorems. All notation is defined and clarified with figures so that the theorems may be readily accessed. Finally, in Chapter 9, a number of practical applications are presented. The examples make apparent the advantages of using integral theorems for engineering analyses. In particular, they provide a set of systematic manipulations for transforming the scale of a problem. Also, arbitrary system geometries are naturally incorporated into the results. Terms that might be inadvertently overlooked or purposefully avoided in standard derivations arise naturally. Furthermore, use of integral theorems provides valuable insight to the physics of a problem. The approach advocated provides very general results which obviates the need to re-derive governing equations every time a small change is made to a system of interest.

1.4 CONCLUSION

The suite of tools presented in this work are helpful for attacking a variety of problems in applied mathematics and engineering. Generalized functions can be used to advantage on problems that require identification of integration regions in space (for example as used to derive the theorems presented here or in obtaining weighted residual statements of finite element methods [*Gray and Celia*, 1990; *Celia and Gray*, 1992]). The integral theorems facilitate the derivation of continuum balance laws at any scale and in transformations of the equations between scales. Additionally, the method of representing vectors and differential operators for arbitrary orthogonal coordinate systems allows problems with arbitrary geometries to be more easily tackled.

1.5 REFERENCES

Anderson, T. B., and R. Jackson, A Fluid Mechanical Description of Fluidized Beds, *Ind. Engr. Chem. Fundamentals*, **6**, 527-539, 1967.

Aris, R., *Vectors, Tensors, and the Basic Equations of Fluid Mechanics*, Prentice-Hall, Englewood Cliffs, NJ, 1962

Bachmat. Y., Spatial Macroscopization of Processes in Heterogeneous Systems, *Israel Journal of Technology*, **10**(5), 391-403, 1972.

Bowen, R. M., *Introduction to Continuum Mechanics for Engineers*, Plenum Press, New York and London, 1989.

Brady, J. F., and G. Bossis, Stokesian Dynamics, *Ann. Rev. Fluid Mech.*, **20**, 111-157, 1988.

Celia, M. A., and W. G. Gray, *Numerical Methods for Differential Equations, Fundamental Concepts for Scientific and Engineering Applications,* Prentice Hall, Englewood Cliffs, NJ, 1992.

Drew, D. A., Averaged Field Equations for Two-Phase Media, *Studies in Appl. Mathematics*, **50**, 133, 1971.

Dull, R. W., *Mathematics for Engineers*, McGraw-Hill, New York, 1926.

Frisch, U., D. d'Humières, B. Hasslacher, P. Lallemand, Y. Pomeau, and J-P. Rivet, *Complex Systems 1*, 648, 1987. (Reprinted in *Lattice Gas Methods for Partial Differential Equations*, edited by G. D. Doolen, Addison-Wesley, 1990.)

Gray, W. G., and M. A. Celia, On the Use of Generalized Functions in Engineering Analysis, *Int. J. Appl. Engng. Ed.*, **6**(1), 89-96, 1990.

Gray, W. G., and P. C. Y. Lee, On the Theorems for Local Volume Averaging of Multiphase Systems, *Int. J. Multiphase Flow*, **3**, 333-340, 1977.

Gray, W. G., and S. M. Hassanizadeh, Averaging Theorems and Averaged Equations for Transport of Interface Properties in Multiphase Systems, *Int. J. Multiphase Flow*, **15**, 81-95, 1989.

Greenberg, M. D., *Foundations of Advanced Mathematics,* Prentice-Hall, Englewood Cliffs, NJ, 1978.

Hassanizadeh, S. M., and W. G. Gray, General Conservation Equations for Multi-phase Systems: 1. Averaging Procedure, *Adv. in Water Resources*, **2**(3), 131-144, 1979

Hassanizadeh, S. M., and W. G. Gray, Mechanics and Thermodynamics of Multiphase Flow in Porous Media Including Interphase Boundaries, *Adv. in Water Resources*, **13**(4), 169-186, 1990.

Hildebrand, F. B., *Advanced Calculus for Applications*, 2nd ed., Prentice-Hall, Englewood Cliffs, NJ, 1976.

Kreyszig, E., *Advanced Engineering Mathematics*, 4th ed., Wiley, New York, 1979.

Malvern, L. E., *Introduction to the Mechanics of a Continuous Medium*, Prentice-Hall, Englewood Cliffs, 1969.

O'Neil, P. V., *Advanced Engineering Mathematics*, 2nd ed., Wadsworth, Belmont, CA, 1987.

Slattery, J. C., Flow of Viscoelastic Fluids through Porous Media, *AIChE Journal*, **13**, 1066, 1967.

Stone, H. A., A simple derivation of the time-dependent convective-diffusion equation for surfactant transport along a deforming interface, *Physics of Fluids A*, **2**(1), 111-112, 1990.

Truesdell, C., *A First Course in Rational Continuum Mechanics*, Academic Press, Boston, 1991.

Whitaker, S., Diffusion and Dispersion in Porous Media, *AIChE Journal*, **13**, 420, 1967.

Wylie, C. R., Jr., *Advanced Engineering Mathematics*, 3rd ed., McGraw-Hill, New York, 1966.

Zill, D. G., and M. R. Cullen, *Advanced Engineering Mathematics*, PWS-Kent, Boston, 1992.

CHAPTER TWO

MATHEMATICAL PRELIMINARIES

2.0 INTRODUCTION

In this book, theorems are derived which transform derivatives from one spatial scale to another. Two classes of theorems are of particular interest: 1) integration theorems and 2) averaging theorems. A transformation of scale is accomplished by integrating smaller scale equations to obtain forms appropriate at larger scales. Three spatial scales of integration will be discussed in the context of the quantitative theorems in Chapter 3. A qualitative description of the scales of interest, designated as microscopic, macroscopic, and megascopic, is provided here. The microscopic scale is defined above the molecular level and is often considered as a continuum scale of variation. The macroscopic scale is an intermediate scale and may be used to model variation of some average property of interest. Transformation to the macroscale perspective may be accomplished by integrating microscale equations and/or properties over some region of space which has a characteristic length much greater than that of the microscale but much smaller than that of the system. At the megascopic scale, variation may be specified from region to region. The megascopic scale is on the order of the length scale of the system under study.

A mathematical framework is established to structure the transformations from one scale to another. The transformation of a time derivative or spatial operator may be applied along a curve, on a surface, or within a volume. Furthermore, for convenience, a set of orthonormal unit vectors, $\boldsymbol{\lambda}$, $\mathbf{n}$, and $\boldsymbol{\nu}$, associated with a curve, a surface, or a volume is set forth as part of the mathematical framework to identify coordinate directions within these regions. For example, $\boldsymbol{\lambda}$ often indicates direction along a curve. The normal direction to a surface or volume is usually given by $\mathbf{n}$, and $\boldsymbol{\nu}$ can be used to represent a coordinate direction within a surface.

Another necessary part of the mathematical framework is the definition of two related coordinate systems, a local and a global coordinate system. The derivation and subsequent application of integration theorems are facilitated by the use of these coordinate systems. Integration regions which conceptualize scale transformations are defined with reference to these coordinate systems. The necessity for and description of the global and local coordinate systems is outlined below in terms of the two theorem classifications being considered.

Integration theorems replace the need to describe at least some of the detailed dependence of the derivative of a function on space with the need to describe the function at the boundary of the domain. These theorems are applied to gradient, divergence, curl, and time derivative operators acting on suitable functions. In the transformation affected by the theorems, functional dependence on one, two, or all three spatial coordinates may be eliminated. Integration over all spatial dependence of a function, causes the function to become completely megascopic, and only values of the function on the boundary and a single averaged value of the function within the system are needed. No information is retained regarding functional dependence on the coordinate directions over which integration has occurred. For integration theorems, where any scale transformation is to the megascale, only a single global, $\mathbf{x}$, coordinate system is needed to identify a point in space. This global coordinate system may have its origin located anywhere in space and is used to identify positions within the integration domain.

Averaging theorems convert microscale equations to equations relevant at some mix of macroscopic and megascopic scales. In the averaging process, derivatives of microscopic functions are transformed to macroscopic derivatives of averaged, macroscopic functions. Integration regions may be curves, surfaces, or volumes; but by employing generalized functions, all integrations may be converted to integrations over volumes. The generalized functions are defined and discussed further in Chapter 3.

Transformation of a derivative to the macroscale or some mix of macro- and megascales making use of averaging theorems, always involves integration over a representative elementary volume (REV). An averaged function and/or property obtained from integration over the REV is associated at the new spatial scale with the location of the REV. Because REV's may be centered at every point in the domain of interest, macroscopically averaged quantities are continuous functions of space. The position of an REV in space is identified through use of the global, $\mathbf{x}$, coordinate system. The origin of the global coordinate system may be arbitrarily located in space. In the context of averaging theorems, the global coordinate system is used to locate positions in the macroscopic or combined macroscopic-megascopic domain. In order to identify points at a smaller scale, within the averaging volume, a second coordinate system, the local, $\boldsymbol{\xi}$, coordinate system is used. The origin of this local coordinate system is at the point used to indicate the location of the averaging volume. The need for a local coordinate system arises when averaging is employed to obtain spatial variation of a macroscopic function (i.e. in fully megascopic averaging or in a direction of megascopic averaging, only one coordinate system need be used). For example, integration theorems involve transformation of functional dependences to the megascale. Thus the integration eliminates functional dependence on the coordinate of integration. Averaging theorems, however, retain functional dependence but shift the dependence from the microscale to the macroscale. Thus macroscale values of a function at a point are really values averaged over a neighborhood of that point. Despite the averaging, every point in the domain of interest will have an averaged value associated with it; and thus it makes sense to consider spatial derivatives of the average values. The use of both local and global coordinate systems facilitates manipulations necessary for derivation of averaging theorems.

In this chapter, the coordinate systems, unit vectors, and derivative operators employed in the derivation and utilization of divergence, transport, and averaging theorems will be detailed. Caution must be exercised in identifying the scale of the coordinates used in differentiating and integrating a function.

2.1 DIFFERENTIATION WITH RESPECT TO GLOBAL OR LOCAL COORDINATES

Sometimes it is necessary to differentiate a function with respect to the global coordinates, $\mathbf{x}$, while at other times, differentiation is with respect to the local coordinates, $\boldsymbol{\xi}$. For example, differentiation of a macroscopically averaged quantity with respect to global coordinates yields an expression for the variation of the averaged quantity throughout space. Differentiation of an averaged quantity with respect to local coordinates is meaningless because local variation is below the scale at which the averaged quantity is known. Throughout the development of the integration and averaging theorems, it is important to be cognizant of the coordinate system being used for spatial differentiation. In particular, the spatial operator may assume different meanings throughout development of the integration theorems. Use of the del operator, ∇, to indicate differentiation with respect to either $\mathbf{x}$ or $\boldsymbol{\xi}$ coordinates must be done carefully to determine whether or not differentiation and integration can be interchanged, a fundamental goal of the averaging process. Thus, it is necessary to define the del operator in terms of both the local and the global coordinate systems.

The position of a point in space, $\mathbf{r}$, can equivalently be written in terms of the location of an averaging volume, $\mathbf{x}$, and the position of the point with respect to that location, $\boldsymbol{\xi}$. Mathematically, this position is expressed as:

$$\mathbf{r} = \mathbf{x} + \boldsymbol{\xi} \tag{2.1}$$

Figure 2.1 demonstrates use of the equivalent position vectors $\mathbf{r}$ and $\mathbf{x} + \boldsymbol{\xi}$ in locating points within an REV. In this figure, the "location" of each REV, $\mathbf{x}$, is selected to be its centroid. A function, f which depends on position in space may be expressed equivalently as $f(\mathbf{r})$ or $f(\mathbf{x} + \boldsymbol{\xi})$. The gradient of this function may be obtained by differentiation using either $\mathbf{r}$ or $\mathbf{x} + \boldsymbol{\xi}$ as the independent variable such that:

$$\nabla_{\mathbf{r}} f = \nabla_{\mathbf{x}+\boldsymbol{\xi}} f \tag{2.2}$$

The differential of f, df, can also be obtained as:

$$df = \nabla_{\mathbf{x}+\boldsymbol{\xi}} f \cdot d(\mathbf{x} + \boldsymbol{\xi}) \tag{2.3}$$

or, by expansion:

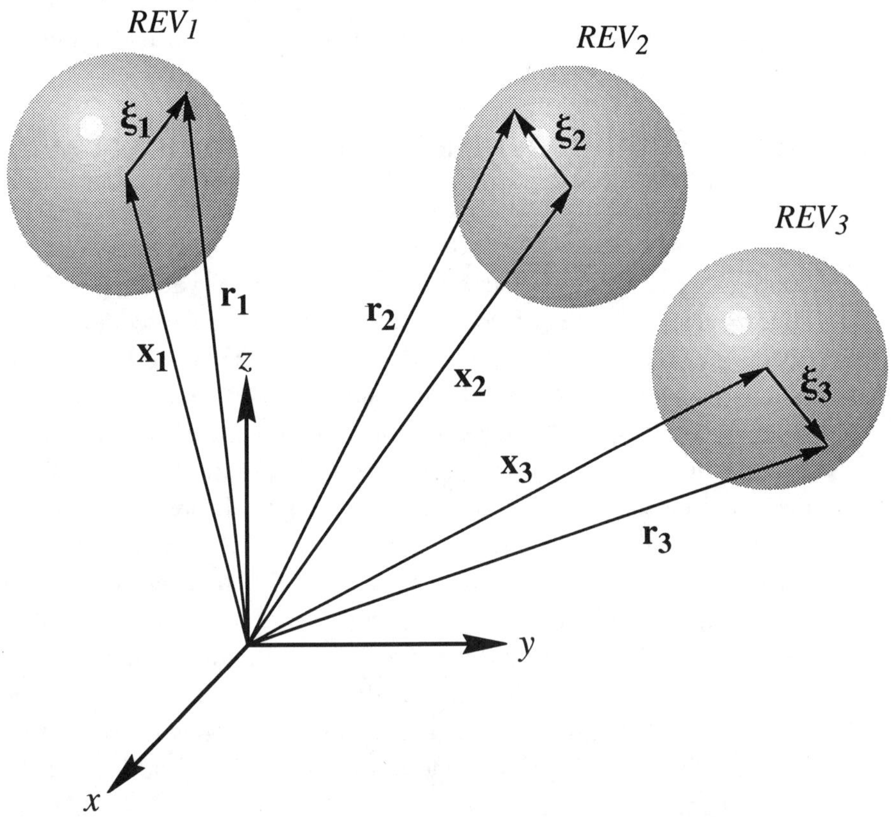

Figure 2.1. Use of the equivalent position vectors $\mathbf{r}$ and $\mathbf{x} + \boldsymbol{\xi}$ to locate a point within an REV. The vector $\mathbf{x}$ locates the centroid of the REV while $\boldsymbol{\xi}$ is the position relative to the centroid.

$$df = \nabla_{\mathbf{x}+\boldsymbol{\xi}} f \cdot d\mathbf{x} + \nabla_{\mathbf{x}+\boldsymbol{\xi}} f \cdot d\boldsymbol{\xi} \quad \textbf{(2.4)}$$

From equation (2.4), the partial derivative of f with respect to the $\mathbf{x}$-coordinates holding the $\boldsymbol{\xi}$-coordinates constant may be derived and expressed:

$$\left.\frac{\partial f}{\partial \mathbf{x}}\right|_{\boldsymbol{\xi}} = \nabla_{\mathbf{x}+\boldsymbol{\xi}} f \quad \textbf{(2.5a)}$$

or:

$$\nabla_{\mathbf{x}} f = \nabla_{\mathbf{x}+\boldsymbol{\xi}} f \quad \textbf{(2.5b)}$$

where $\nabla_{\mathbf{x}}$ is the gradient operator with respect to $\mathbf{x}$ coordinates holding $\boldsymbol{\xi}$ coordinates constant. In a similar fashion, the partial derivative of f with respect to the $\boldsymbol{\xi}$ coordinates holding the $\mathbf{x}$ coordinates constant is written:

$$\left.\frac{\partial f}{\partial \boldsymbol{\xi}}\right|_{\mathbf{x}} = \nabla_{\mathbf{x}+\boldsymbol{\xi}} f \quad \textbf{(2.6a)}$$

or in the more convenient form:

$$\nabla_{\boldsymbol{\xi}} f = \nabla_{\mathbf{x}+\boldsymbol{\xi}} f \quad \textbf{(2.6b)}$$

From equations (2.2), (2.5b), and (2.6b), the following relationship is derived when $f(\mathbf{r})$ is equivalent to $f(\mathbf{x}+\boldsymbol{\xi})$:

$$\nabla_{\mathbf{r}} f = \nabla_{\mathbf{x}+\boldsymbol{\xi}} f = \nabla_{\mathbf{x}} f = \nabla_{\boldsymbol{\xi}} f \quad \textbf{(2.7)}$$

Thus for a function whose spatial dependence may be expressed in terms of $\mathbf{x}+\boldsymbol{\xi}$ rather than $\mathbf{x}$ and $\boldsymbol{\xi}$ separately, the gradient with respect to either the $\mathbf{x}$ or the $\boldsymbol{\xi}$ coordinate is identical as given by equation (2.7). For simplicity, the explicit distinction among the various forms of the gradient operators often will not be made. Rather the operator ∇ will appear without a subscript, and the coordinate system to which the spatial operator is referenced can be inferred from the resulting theorem.

2.2 CONVENTIONS FOR ORTHONORMAL VECTORS

The mathematical framework outlined thus far allows the freedom to select the shape and orientation of an averaging volume. However, surfaces and curves in space that correspond to physical systems are not necessarily regular in shape. Thus, a set of orthogonal vectors is defined which is used to describe arbitrary surfaces and curves in space. The vectors designated $\boldsymbol{\lambda}$, $\mathbf{n}$, and $\boldsymbol{\nu}$ constitute a right-handed triple of orthogonal unit vectors. Figure 2.2 illustrates the relationship among these vectors.

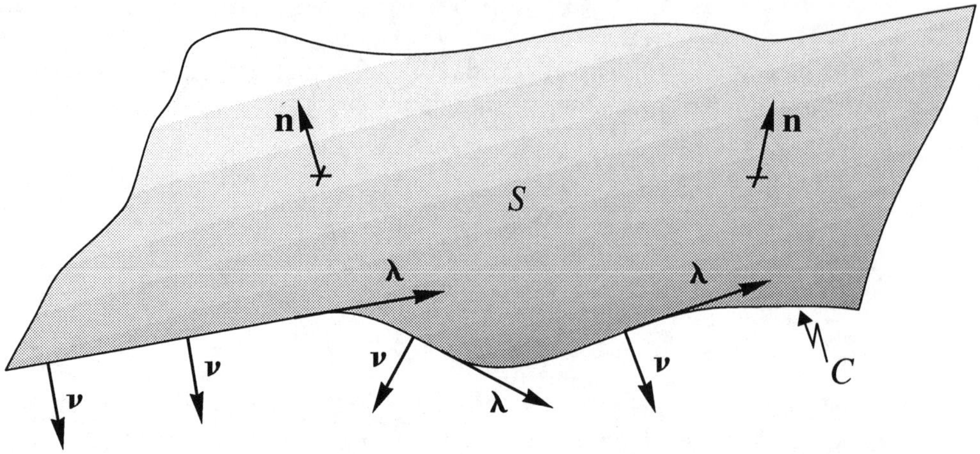

Figure 2.2. A surface S with a portion of its boundary curve designated as C illustratrating the orthonormal set of vectors $\boldsymbol{\lambda}$, $\mathbf{n}$, and $\boldsymbol{\nu}$. Here $\boldsymbol{\lambda}$ is tangent to C, $\mathbf{n}$ is normal to S, and $\boldsymbol{\nu}$ is not only tangent to S but also normal to C.

To ensure that $\boldsymbol{\lambda}$, $\mathbf{n}$, and $\boldsymbol{\nu}$ will be uniquely defined, only orientable surfaces and simple curves are considered herein. A surface is orientable if the positive normal direction, indicated by the vector $\mathbf{n}$ in figure 2.2, can be defined in a unique and continuous way around the surface [*Kreyszig*, 1962]. Physically an orientable surface such as S in figure 2.2 has two distinguishable sides which permit selection of a positive normal direction. A well known example of a non-orientable surface is the Möbius strip (A Möbius strip may be constructed by taking a long flat piece of rectangular paper with sides indicated as "a" and "b", rotating one of the narrow ends 180°, and joining the narrow ends of the paper to form a loop by overlapping the ends such that side "a" is the only exposed side at the point of joining.). For this surface, the positive normal direction reverses when the normal vector is displaced continuously along the surface from a starting point, around the loop, back to the starting point. Of course, when cut, the Möbius strip becomes an orientable surface. So, for any orientable surface, the normal vector to the surface (i.e. $\mathbf{n}$) uniquely defines that surface. Note that one degree of arbitrariness exists in that two choices exist for the positive normal direction. Provided one is consistent with manipulations, this degree of freedom is unimportant. Here, for a closed surface, $\mathbf{n}$ will be selected as positive outward. The vectors $\boldsymbol{\nu}$ and $\boldsymbol{\lambda}$ are not uniquely defined for a surface, though they are constrained to be orthogonal to $\mathbf{n}$ and to each other.

Curves over which the averaging procedure is valid must be simple curves such as the curve C in figure 2.2. Simple curves have no points of self-intersection with the consequence that at a point, one tangent vector, $\boldsymbol{\lambda}$, orients the curve. Here, as with the discussion for surfaces, two choices exist for selection of the positive $\boldsymbol{\lambda}$ direction; but this does not affect the problem. Integration along non-simple curves is accomplished by defining them as piecewise simple curves (i.e. by cutting them into pieces that are simple curves). Once the positive direction along the curve is selected, the tangential unit vector $\boldsymbol{\lambda}$ is unique and varies continuously along C. Along the same curve, unit vectors $\mathbf{n}$ and $\boldsymbol{\nu}$ are not necessarily uniquely defined. They are restricted to being orthogonal to $\boldsymbol{\lambda}$ and to each other, but neither is required here to coincide with the direction of principal curvature. If, however, the curve is the edge of a surface, as in figure 2.2, the directions of all three unit vectors are unique with $\mathbf{n}$ normal to the surface, $\boldsymbol{\lambda}$ tangential to the bounding curve and normal to $\mathbf{n}$, and $\boldsymbol{\nu}$ orthogonal to both $\mathbf{n}$ and $\boldsymbol{\lambda}$. The coordinate system is usually selected so that $\boldsymbol{\nu}$ is outwardly directed from the surface.

A few useful relationships among the orthogonal unit vectors, $\boldsymbol{\lambda}$, $\mathbf{n}$, and $\boldsymbol{\nu}$, are presented below without proof. These relationships become very useful in deriving expressions for the spatial and time derivatives of functions in subsequent sections. First, the relations that define the vectors as orthogonal and of unit magnitude are:

$$\boldsymbol{\lambda}\cdot\boldsymbol{\lambda} = \mathbf{n}\cdot\mathbf{n} = \boldsymbol{\nu}\cdot\boldsymbol{\nu} = 1 \tag{2.8}$$

$$\boldsymbol{\lambda}\cdot\mathbf{n} = \boldsymbol{\lambda}\cdot\boldsymbol{\nu} = \mathbf{n}\cdot\boldsymbol{\nu} = 0 \tag{2.9}$$

$$\boldsymbol{\lambda}\times\mathbf{n} = \boldsymbol{\nu} \tag{2.10a}$$

$$\mathbf{n}\times\boldsymbol{\nu} = \boldsymbol{\lambda} \tag{2.10b}$$

$$\boldsymbol{\nu}\times\boldsymbol{\lambda} = \mathbf{n} \tag{2.10c}$$

$$\boldsymbol{\lambda}\times\boldsymbol{\lambda} = \mathbf{n}\times\mathbf{n} = \boldsymbol{\nu}\times\boldsymbol{\nu} = 0 \tag{2.11}$$

The relations among the derivatives of these unit vectors are then obtained as:

$$\nabla\boldsymbol{\lambda}\cdot\boldsymbol{\lambda} = \nabla\mathbf{n}\cdot\mathbf{n} = \nabla\boldsymbol{\nu}\cdot\boldsymbol{\nu} = 0 \tag{2.12}$$

$$\frac{\partial\boldsymbol{\lambda}}{\partial t}\cdot\boldsymbol{\lambda} = \frac{\partial\mathbf{n}}{\partial t}\cdot\mathbf{n} = \frac{\partial\boldsymbol{\nu}}{\partial t}\cdot\boldsymbol{\nu} = 0 \tag{2.13}$$

$$\nabla\mathbf{n}\cdot\boldsymbol{\nu} = -\nabla\boldsymbol{\nu}\cdot\mathbf{n} \tag{2.14a}$$

$$\nabla\boldsymbol{\nu}\cdot\boldsymbol{\lambda} = -\nabla\boldsymbol{\lambda}\cdot\boldsymbol{\nu} \tag{2.14b}$$

$$\nabla\boldsymbol{\lambda}\cdot\mathbf{n} = -\nabla\mathbf{n}\cdot\boldsymbol{\lambda} \tag{2.14c}$$

$$\nabla\times\boldsymbol{\lambda} = (\boldsymbol{\lambda}\boldsymbol{\lambda}\cdot\nabla)\times\boldsymbol{\lambda} = -\boldsymbol{\lambda}\cdot\nabla\boldsymbol{\lambda}\times\boldsymbol{\lambda} \tag{2.15a}$$

$$\nabla\times\mathbf{n} = (\mathbf{n}\mathbf{n}\cdot\nabla)\times\mathbf{n} = -\mathbf{n}\cdot\nabla\mathbf{n}\times\mathbf{n} \tag{2.15b}$$

$$\nabla\times\boldsymbol{\nu} = (\boldsymbol{\nu}\boldsymbol{\nu}\cdot\nabla)\times\boldsymbol{\nu} = -\boldsymbol{\nu}\cdot\nabla\boldsymbol{\nu}\times\boldsymbol{\nu} \tag{2.15c}$$

The unit tensor in terms of the orthonormal vectors may be written:

$$\mathbf{I} = \boldsymbol{\lambda}\boldsymbol{\lambda} + \mathbf{n}\mathbf{n} + \boldsymbol{\nu}\boldsymbol{\nu} \tag{2.16}$$

The above identities are also used throughout derivation of the integration theorems in later chapters.

2.3 SPATIAL DERIVATIVES

2.3.1 Motivation for the Definition of General Spatial Derivatives

Determination of the gradient, divergence, or curl of a function contained within a surface requires definition of a surface del operator. Likewise when considering the divergence, gradient, of a function defined only along a curve, a one-dimensional del operator is necessary. The surficial operator, written ∇^s, and the operator along a curve, expressed ∇^c, are therefore required. Spatial derivatives referenced to a particular region, such as a volume, a surface, or a curve, are intrinsic to the development of the theorems presented in this manuscript.

The orientation of unit vectors associated with a cartesian coordinate system is independent of position. Thus for a cartesian system, the spatial del operator may be defined as:

$$\nabla = \boldsymbol{\lambda}\frac{\partial}{\partial \lambda} + \mathbf{n}\frac{\partial}{\partial n} + \boldsymbol{\nu}\frac{\partial}{\partial \nu} \tag{2.17}$$

where the unit vectors are parallel to each of the three cartesian coordinate directions (λ, n, ν). However for other coordinate systems, such as cylindrical and spherical, the orientation of the unit vectors depends on position. Derivatives of these orthogonal vectors, often referred to as scale factors [*Hildebrand*, 1976], arise from rotations and/or translations of the orthogonal coordinate vectors in space. For the general spatial operators used in this work, these scale factors are not explicitly expressed but are embedded within the definitions of the general spatial operators for a volume, a surface, and a curve.

Averaging involves integration over curves, surfaces, or volumes contained within a region of space. By the approach adopted here, relations among integrations over these different types of regions are invoked. In addition, relations among the three-dimensional spatial operator, ∇, the two-dimensional spatial operator, ∇^s, and the one-dimensional spatial operator, ∇^c are employed. The orthogonal unit vectors, $\boldsymbol{\lambda}$, $\mathbf{n}$, and $\boldsymbol{\nu}$, facilitate the development of these interrelations without need for reliance on the conceptually difficult concepts of covariant and contravariant differentiation [e.g. as in *Aris*, 1962; *Moeckel*, 1975].

2.3.2 Definition of One-, Two-, and Three-Dimensional Spatial Derivatives

A vector function, $\mathbf{f}$, can be written in component form as:

$$\mathbf{f} = \boldsymbol{\lambda\lambda}\bullet\mathbf{f} + \mathbf{nn}\bullet\mathbf{f} + \boldsymbol{\nu\nu}\bullet\mathbf{f} \tag{2.18}$$

where $\boldsymbol{\lambda}$, $\mathbf{n}$, and $\boldsymbol{\nu}$ are the previously defined unit vectors for an arbitrary orthogonal coordinate system. The terms $\boldsymbol{\lambda}\bullet\mathbf{f}$, $\mathbf{n}\bullet\mathbf{f}$, and $\boldsymbol{\nu}\bullet\mathbf{f}$ are the components of $\mathbf{f}$ in the $\boldsymbol{\lambda}$, $\mathbf{n}$, and $\boldsymbol{\nu}$ directions respectively. Therefore if one wishes to consider only the vector components of $\mathbf{f}$ in a surface with unit normal $\mathbf{n}$, the surface vector, $\mathbf{f}^s$, may be obtained as:

$$\mathbf{f}^s = \mathbf{f} - \mathbf{nn}\bullet\mathbf{f} \tag{2.19}$$

Similarly, the vector component of $\mathbf{f}$ along a curve with tangent $\boldsymbol{\lambda}$ is given by the vector $\mathbf{f}^c$ where:

$$\mathbf{f}^c = \boldsymbol{\lambda\lambda}\bullet\mathbf{f} \tag{2.20}$$

Now define a vector, $\mathbf{h}$, in terms of the gradient of a scalar, g, as:

$$\mathbf{h} = \nabla g \tag{2.21}$$

The vector expansion of $\mathbf{h}$ follows that of equation (2.18) and is written as:

$$\mathbf{h} = \boldsymbol{\lambda\lambda}\bullet\mathbf{h} + \mathbf{nn}\bullet\mathbf{h} + \boldsymbol{\nu\nu}\bullet\mathbf{h} \tag{2.22}$$

Substitution of equation (2.21) into equation (2.22) yields:

$$\mathbf{h} = \boldsymbol{\lambda\lambda}\bullet\nabla g + \mathbf{nn}\bullet\nabla g + \boldsymbol{\nu\nu}\bullet\nabla g \tag{2.23}$$

which can also be expressed as:

$$\mathbf{h} = (\boldsymbol{\lambda\lambda}\bullet\nabla + \mathbf{nn}\bullet\nabla + \boldsymbol{\nu\nu}\bullet\nabla)\, g \tag{2.24}$$

Comparison of equation (2.21) with equation (2.24) indicates that the three-dimensional spatial del operator may be expressed as follows:

$$\nabla = \boldsymbol{\lambda\lambda}\bullet\nabla + \mathbf{nn}\bullet\nabla + \boldsymbol{\nu\nu}\bullet\nabla \tag{2.25}$$

This expression provides the gradient, divergence, or curl operator in its component form. For a cartesian coordinate system, equations (2.25) and (2.17) are identical.

Since a surface is uniquely defined by its unit normal vector, the surficial operator is expressed in terms of the three-dimensional del operator, ∇, and the unit vector normal to the surface, indicated as the vector $\mathbf{n}$ in figure 2.2. The surficial form of the del vector operator is equal to the spatial form minus the component normal to the surface or:

$$\nabla^s = \nabla - \mathbf{n}\mathbf{n}\cdot\nabla \tag{2.26}$$

A curve is uniquely defined by specification of its tangent vector. The curvilineal form (i.e. form along a curve) of the del operator along the curve C in figure 2.2 with tangent $\boldsymbol{\lambda}$ is simply the component of the spatial operator in that direction:

$$\nabla^c = \boldsymbol{\lambda}\boldsymbol{\lambda}\cdot\nabla \tag{2.27}$$

An indication of the generality and versatility of the spatial operators just set forth is seen in the subsequent subsections through brief examination of the gradient, divergence, and curl of spatial, surficial, and curvilineal quantities.

2.3.3 Gradient of a Scalar Function

The expression for the spatial gradient of a function may be obtained in terms of its component parts by employing equation (2.25) as:

$$\nabla f = \boldsymbol{\lambda}\boldsymbol{\lambda}\cdot\nabla f + \mathbf{n}\mathbf{n}\cdot\nabla f + \boldsymbol{\nu}\boldsymbol{\nu}\cdot\nabla f \tag{2.28}$$

It will sometimes prove useful to be able to calculate the gradient of a function within a surface. If $\boldsymbol{\lambda}$ is a unit vector normal to the surface, the surface gradient of a function may be obtained as:

$$\nabla^s f = \mathbf{n}\mathbf{n}\cdot\nabla f + \boldsymbol{\nu}\boldsymbol{\nu}\cdot\nabla f \tag{2.29}$$

Note that in defining the surface gradient, it is important to explicitly designate the unit vector that will be selected as normal to the surface. For example, the surface gradient that would be calculated using the operator defined in equation (2.26) would be that for a surface with unit normal **n** and would have the form:

$$\nabla^s f = \boldsymbol{\lambda}\boldsymbol{\lambda}\cdot\nabla f + \boldsymbol{\nu}\boldsymbol{\nu}\cdot\nabla f \tag{2.30}$$

The gradient along a curve may also be expressed in terms of the component of the spatial gradient operator tangent to the curve. If the unit vector tangent to a curve is $\boldsymbol{\lambda}$, the curvilineal gradient is:

$$\nabla^c f = \boldsymbol{\lambda}\boldsymbol{\lambda}\cdot\nabla f \tag{2.31}$$

A convenient relation among the spatial, surficial, and curvilineal gradients is obtained by substitution of equations (2.29) and (2.31) into equation (2.28) to obtain:

$$\nabla f = \nabla^s f + \nabla^c f \tag{2.32}$$

This relation is valid for the case of a curve C whose tangent is normal to the surface S. Particular forms that facilitate the evaluation of the gradient in cartesian, cylindrical, and spherical coordinates are provided in Table 2.1.

2.3.4 Divergence of a Spatial Vector

An expression for the divergence of a spatial vector, $\mathbf{f}$, may be obtained such that the orthogonal unit vectors, $\boldsymbol{\lambda}$, $\mathbf{n}$, and $\boldsymbol{\nu}$, appear explicitly. This relation has some utility in the derivation of the averaging theorems in subsequent sections. The derivation of $\nabla \bullet \mathbf{f}$ is obtained by using the definitions of ∇ and $\mathbf{f}$ given in equations (2.25) and (2.18) respectively to obtain:

$$\nabla \bullet \mathbf{f} = (\boldsymbol{\lambda\lambda} \bullet \nabla + \mathbf{nn} \bullet \nabla + \boldsymbol{\nu\nu} \bullet \nabla) \bullet (\boldsymbol{\lambda\lambda} \bullet \mathbf{f} + \mathbf{nn} \bullet \mathbf{f} + \boldsymbol{\nu\nu} \bullet \mathbf{f}) \tag{2.33}$$

This expression may be manipulated rather easily using indicial notation. However, rather than introduce that technique here, the vector notation will be retained. Expansion of equation (2.33) results in nine terms such that:

$$\begin{aligned}\nabla \bullet \mathbf{f} = {} & (\boldsymbol{\lambda\lambda} \bullet \nabla) \bullet (\boldsymbol{\lambda\lambda} \bullet \mathbf{f}) + (\boldsymbol{\lambda\lambda} \bullet \nabla) \bullet (\mathbf{nn} \bullet \mathbf{f}) + (\boldsymbol{\lambda\lambda} \bullet \nabla) \bullet (\boldsymbol{\nu\nu} \bullet \mathbf{f}) \\ & + (\mathbf{nn} \bullet \nabla) \bullet (\boldsymbol{\lambda\lambda} \bullet \mathbf{f}) + (\mathbf{nn} \bullet \nabla) \bullet (\mathbf{nn} \bullet \mathbf{f}) + (\mathbf{nn} \bullet \nabla) \bullet (\boldsymbol{\nu\nu} \bullet \mathbf{f}) \\ & + (\boldsymbol{\nu\nu} \bullet \nabla) \bullet (\boldsymbol{\lambda\lambda} \bullet \mathbf{f}) + (\boldsymbol{\nu\nu} \bullet \nabla) \bullet (\mathbf{nn} \bullet \mathbf{f}) + (\boldsymbol{\nu\nu} \bullet \nabla) \bullet (\boldsymbol{\nu\nu} \bullet \mathbf{f})\end{aligned} \tag{2.34}$$

The relations on the right side of equation (2.34) may be simplified making use of the fact that the properties of orthogonal vectors stated in relations (2.8), (2.9), and (2.12) apply to obtain:

$$\begin{aligned}\nabla \bullet \mathbf{f} = {} & \boldsymbol{\lambda} \bullet \nabla (\boldsymbol{\lambda} \bullet \mathbf{f}) + \boldsymbol{\lambda} \bullet \nabla \mathbf{n} \bullet \boldsymbol{\lambda} \mathbf{n} \bullet \mathbf{f} + \boldsymbol{\lambda} \bullet \nabla \boldsymbol{\nu} \bullet \boldsymbol{\lambda} \boldsymbol{\nu} \bullet \mathbf{f} \\ & + \mathbf{n} \bullet \nabla \boldsymbol{\lambda} \bullet \mathbf{n} \boldsymbol{\lambda} \bullet \mathbf{f} + \mathbf{n} \bullet \nabla (\mathbf{n} \bullet \mathbf{f}) + \mathbf{n} \bullet \nabla \boldsymbol{\nu} \bullet \mathbf{n} \boldsymbol{\nu} \bullet \mathbf{f} \\ & + \boldsymbol{\nu} \bullet \nabla \boldsymbol{\lambda} \bullet \boldsymbol{\nu} \boldsymbol{\lambda} \bullet \mathbf{f} + \boldsymbol{\nu} \bullet \nabla \mathbf{n} \bullet \boldsymbol{\nu} \mathbf{n} \bullet \mathbf{f} + \boldsymbol{\nu} \bullet \nabla (\boldsymbol{\nu} \bullet \mathbf{f})\end{aligned} \tag{2.35}$$

Equations (2.14a), (2.14b), and (2.14c) can be used to rearrange the six terms in this equation that involve the gradient of a unit vector. Application of these relations and collection of terms in each row of the equation yield:

$$\begin{aligned}\nabla \bullet \mathbf{f} = {} & \boldsymbol{\lambda} \bullet \nabla (\boldsymbol{\lambda} \bullet \mathbf{f}) - \boldsymbol{\lambda} \bullet \nabla \boldsymbol{\lambda} \bullet (\mathbf{nn} \bullet \mathbf{f} + \boldsymbol{\nu\nu} \bullet \mathbf{f}) \\ & + \mathbf{n} \bullet \nabla (\mathbf{n} \bullet \mathbf{f}) - \mathbf{n} \bullet \nabla \mathbf{n} \bullet (\boldsymbol{\nu\nu} \bullet \mathbf{f} + \boldsymbol{\lambda\lambda} \bullet \mathbf{f}) \\ & + \boldsymbol{\nu} \bullet \nabla (\boldsymbol{\nu} \bullet \mathbf{f}) - \boldsymbol{\nu} \bullet \nabla \boldsymbol{\nu} \bullet (\boldsymbol{\lambda\lambda} \bullet \mathbf{f} + \mathbf{nn} \bullet \mathbf{f})\end{aligned} \tag{2.36}$$

Substitution of equation (2.18) into the last term of each row of this equation leads to:

Table 2.1: Geometric terms for gradient evaluation in standard coordinate systems

Coordinate System	$\boldsymbol{\lambda}$	$\mathbf{n}$	$\boldsymbol{\nu}$	$\boldsymbol{\lambda}\bullet\nabla^{\lambda}f$	$\mathbf{n}\bullet\nabla^{n}f$	$\boldsymbol{\nu}\bullet\nabla^{\nu}f$
Cartesian	$\mathbf{i}$	$\mathbf{j}$	$\mathbf{k}$	$\frac{\partial f}{\partial x}$	$\frac{\partial f}{\partial y}$	$\frac{\partial f}{\partial z}$
Cylindrical	$\mathbf{e}_r$	$\mathbf{e}_\theta$	$\mathbf{e}_z$	$\frac{\partial f}{\partial r}$	$\frac{1}{r}\frac{\partial f}{\partial \theta}$	$\frac{\partial f}{\partial z}$
Spherical	$\mathbf{e}_r$	$\mathbf{e}_\theta$	$\mathbf{e}_\phi$	$\frac{\partial f}{\partial r}$	$\frac{1}{r}\frac{\partial f}{\partial \theta}$	$\frac{1}{r\sin\theta}\frac{\partial f}{\partial \phi}$

$$\begin{aligned}\nabla\cdot\mathbf{f} &= \boldsymbol{\lambda}\cdot\nabla(\boldsymbol{\lambda}\cdot\mathbf{f}) - \boldsymbol{\lambda}\cdot\nabla\boldsymbol{\lambda}\cdot(\mathbf{f}-\boldsymbol{\lambda}\boldsymbol{\lambda}\cdot\mathbf{f})\\ &\quad+\mathbf{n}\cdot\nabla(\mathbf{n}\cdot\mathbf{f}) - \mathbf{n}\cdot\nabla\mathbf{n}\cdot(\mathbf{f}-\mathbf{n}\mathbf{n}\cdot\mathbf{f})\\ &\quad+\boldsymbol{\nu}\cdot\nabla(\boldsymbol{\nu}\cdot\mathbf{f}) - \boldsymbol{\nu}\cdot\nabla\boldsymbol{\nu}\cdot(\mathbf{f}-\boldsymbol{\nu}\boldsymbol{\nu}\cdot\mathbf{f})\end{aligned} \quad \textbf{(2.37)}$$

Application of the chain rule as well as the relations given in equation (2.12) results in the final expression for the spatial divergence of a spatial vector:

$$\nabla\cdot\mathbf{f} = \boldsymbol{\lambda}\cdot\nabla\mathbf{f}\cdot\boldsymbol{\lambda} + \mathbf{n}\cdot\nabla\mathbf{f}\cdot\mathbf{n} + \boldsymbol{\nu}\cdot\nabla\mathbf{f}\cdot\boldsymbol{\nu} \quad \textbf{(2.38)}$$

Equation (2.38) might be considered an obvious application of equation (2.25) to obtain the divergence of a vector. Indeed, when both the operator and the vector are spatial quantities, this result may be obtained directly. However, when the divergence operator is of reduced dimension, such straightforward application is not possible. In those cases, examination of the operator and vector in component form provides some useful identities.

2.3.5 Surface Divergence of a Surface Vector

In some instances it is necessary to relate the spatial divergence of a spatial vector, $\nabla\cdot\mathbf{f}$ to the surface divergence of the surface components of that vector, $\nabla^s\cdot\mathbf{f}^s$. In deriving this relationship, let $\mathbf{n}$ be the unit vector normal to the surface. Equation (2.19) relates $\mathbf{f}$ to $\mathbf{f}^s$, and equation (2.26) relates ∇ to ∇^s such that:

$$\nabla^s\cdot\mathbf{f}^s = (\nabla - \mathbf{n}\mathbf{n}\cdot\nabla)\cdot(\mathbf{f}-\mathbf{n}\mathbf{n}\cdot\mathbf{f}) \quad \textbf{(2.39)}$$

This equation may be expanded out making use of equation (2.12) which indicates that $\nabla\mathbf{n}\cdot\mathbf{n} = 0$ to obtain:

$$\nabla^s\cdot\mathbf{f}^s = \nabla\cdot\mathbf{f} - \mathbf{n}\cdot\nabla\mathbf{f}\cdot\mathbf{n} - \mathbf{n}\cdot\mathbf{f}\nabla\cdot\mathbf{n} - \mathbf{n}\cdot\nabla(\mathbf{n}\cdot\mathbf{f}) + \mathbf{n}\cdot\nabla(\mathbf{n}\cdot\mathbf{f}) \quad \textbf{(2.40)}$$

The last two terms in this expression cancel, and the second and third terms on the right side may be combined to give the relation:

$$\nabla^s\cdot\mathbf{f}^s = \nabla\cdot\mathbf{f} - \nabla\cdot(\mathbf{n}\mathbf{f})\cdot\mathbf{n} \quad \textbf{(2.41)}$$

This equation also appears in *Moeckel* [1975] and *Gray and Hassanizadeh* [1989]. The mean or Gaussian curvature of a surface, K_M, is calculated as:

$$2K_M = -\nabla\cdot\mathbf{n} \quad \textbf{(2.42)}$$

When this expression is employed, equation (2.41) may alternatively be written:

$$\nabla^s\cdot\mathbf{f}^s = \nabla\cdot\mathbf{f} - \mathbf{n}\cdot\nabla\mathbf{f}\cdot\mathbf{n} + 2K_M\mathbf{f}\cdot\mathbf{n} \quad \textbf{(2.43)}$$

2.3.6 Curvilineal Divergence of a Curvilineal Vector

The expression for the divergence along a curve, ∇^c, of a vector tangent to the curve, $\mathbf{f}^c$, may be related to the spatial divergence of a spatial vector. If the unit vector tangent to the curve is designated as $\boldsymbol{\lambda}$, equations (2.20) and (2.27) may be employed to write:

$$\nabla^c \cdot \mathbf{f}^c = (\boldsymbol{\lambda}\boldsymbol{\lambda} \cdot \nabla) \cdot (\boldsymbol{\lambda}\boldsymbol{\lambda} \cdot \mathbf{f}) \qquad \textbf{(2.44)}$$

Equation (2.12) indicates that $\nabla\boldsymbol{\lambda} \cdot \boldsymbol{\lambda} = 0$, so this equation simplifies to:

$$\nabla^c \cdot \mathbf{f}^c = \boldsymbol{\lambda} \cdot \nabla (\boldsymbol{\lambda} \cdot \mathbf{f}) \qquad \textbf{(2.45)}$$

2.3.7 Relationships among Spatial, Surficial, and Curvilineal Divergences

In performing the integrations over volumes, surfaces, and curves, it is useful to have access to relationships among ∇, ∇^s, and ∇^c as given by equations (2.38), (2.41), and (2.45). In obtaining these relations, appropriate unit vectors must be used to avoid confusion.

First obtain the relationship between the spatial divergence and the sum of the curvilineal divergences taken in each of the three orthogonal directions. In the $\boldsymbol{\lambda}$, $\mathbf{n}$, and $\boldsymbol{\nu}$ directions, the curvilineal divergences are forms of equation (2.45) appropriate for each direction and are given, respectively, by:

$$\nabla^\lambda \cdot \mathbf{f}^\lambda = \boldsymbol{\lambda} \cdot \nabla (\boldsymbol{\lambda} \cdot \mathbf{f}) = \nabla \cdot (\boldsymbol{\lambda}\boldsymbol{\lambda} \cdot \mathbf{f}) - \boldsymbol{\lambda} \cdot \mathbf{f} \nabla \cdot \boldsymbol{\lambda} \qquad \textbf{(2.46a)}$$

$$\nabla^n \cdot \mathbf{f}^n = \mathbf{n} \cdot \nabla (\mathbf{n} \cdot \mathbf{f}) = \nabla \cdot (\mathbf{n}\mathbf{n} \cdot \mathbf{f}) - \mathbf{n} \cdot \mathbf{f} \nabla \cdot \mathbf{n} \qquad \textbf{(2.46b)}$$

$$\nabla^\nu \cdot \mathbf{f}^\nu = \boldsymbol{\nu} \cdot \nabla (\boldsymbol{\nu} \cdot \mathbf{f}) = \nabla \cdot (\boldsymbol{\nu}\boldsymbol{\nu} \cdot \mathbf{f}) - \boldsymbol{\nu} \cdot \mathbf{f} \nabla \cdot \boldsymbol{\nu} \qquad \textbf{(2.46c)}$$

In each of these equations, the terms following the second equal sign are obtained by application of the chain rule to the term after the first equal sign. Summation of these three equations and invocation of the expanded definition of $\mathbf{f}$ given by equation (2.18) yields:

$$\nabla^\lambda \cdot \mathbf{f}^\lambda + \nabla^n \cdot \mathbf{f}^n + \nabla^\nu \cdot \mathbf{f}^\nu = \nabla \cdot \mathbf{f} - \boldsymbol{\lambda} \cdot \mathbf{f} \nabla \cdot \boldsymbol{\lambda} - \mathbf{n} \cdot \mathbf{f} \nabla \cdot \mathbf{n} - \boldsymbol{\nu} \cdot \mathbf{f} \nabla \cdot \boldsymbol{\nu} \qquad \textbf{(2.47a)}$$

or, after rearrangement:

$$\nabla \cdot \mathbf{f} = \nabla^\lambda \cdot \mathbf{f}^\lambda + \nabla^n \cdot \mathbf{f}^n + \nabla^\nu \cdot \mathbf{f}^\nu + \boldsymbol{\lambda} \cdot \mathbf{f} \nabla \cdot \boldsymbol{\lambda} + \mathbf{n} \cdot \mathbf{f} \nabla \cdot \mathbf{n} + \boldsymbol{\nu} \cdot \mathbf{f} \nabla \cdot \boldsymbol{\nu} \qquad \textbf{(2.47b)}$$

Note that the last three terms in equation (2.47b) account for the curvature of each of the three orthogonal coordinates. When $\boldsymbol{\lambda}$, $\mathbf{n}$, and $\boldsymbol{\nu}$ correspond to the unit vectors in a cartesian system, these last terms are each zero. An equivalent form of equation (2.47b) may be obtained by application of the chain rule to the last three terms to obtain:

$$\nabla\cdot\mathbf{f} = \nabla^{\lambda}\cdot\mathbf{f}^{\lambda} + \nabla^{n}\cdot\mathbf{f}^{n} + \nabla^{\nu}\cdot\mathbf{f}^{\nu} - \boldsymbol{\lambda}\cdot\nabla\boldsymbol{\lambda}\cdot\mathbf{f} - \mathbf{n}\cdot\nabla\mathbf{n}\cdot\mathbf{f} - \boldsymbol{\nu}\cdot\nabla\boldsymbol{\nu}\cdot\mathbf{f} \quad \textbf{(2.47c)}$$

where use has been made of the fact that the divergence of the unit tensor, $\mathbf{I}$, as given by equation (2.16), is zero.

A second equation of interest relates the sum of the surface divergence and the curvilineal divergence to the spatial divergence. For this expression, the normal to the surface is the curve along which the curvilineal divergence is evaluated (see figure 2.3). If this direction is identified as $\mathbf{n}$, then $\boldsymbol{\lambda}$ in equation (2.45) must be replaced by $\mathbf{n}$. Summation of equation (2.41) with this form of equation (2.45) yields:

$$\nabla^{s}\cdot\mathbf{f}^{s} + \nabla^{c}\cdot\mathbf{f}^{c} = \nabla\cdot\mathbf{f} - \nabla\cdot(\mathbf{nf})\cdot\mathbf{n} + \mathbf{n}\cdot\nabla(\mathbf{n}\cdot\mathbf{f}) \quad \textbf{(2.48)}$$

Application of the chain rule to the last two terms in this equation and rearrangement to obtain an expression for $\nabla\cdot\mathbf{f}$ leads to the identity:

$$\nabla\cdot\mathbf{f} = \nabla^{s}\cdot\mathbf{f}^{s} + \nabla^{c}\cdot\mathbf{f}^{c} + \mathbf{n}\cdot\mathbf{f}\nabla\cdot\mathbf{n} - \mathbf{n}\cdot\nabla\mathbf{n}\cdot\mathbf{f} \quad \textbf{(2.49)}$$

The last two terms in this expression account for the curvature of the coordinate normal to the surface in the direction of the curve. For a cartesian coordinate system, these terms are zero.

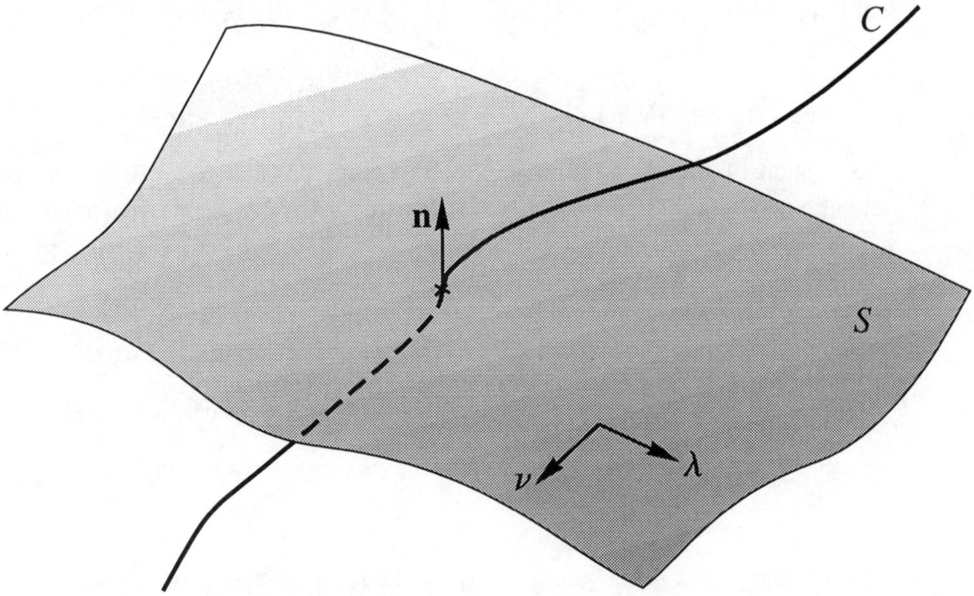

Figure 2.3. A curve C intersecting surface S such that it is normal to S. For this case, $\mathbf{n}$ is normal to S and tangent to C. The orthogonal coordinate directions λ and ν are tangent to the surface S.

A final set of identities that is of interest relates a surficial divergence to a curvilineal divergence when the curve lies in the surface. The derivation of this expression is similar to those above. With reference to figure 2.3, let λ and ν be the coordinates in the surface with unit vectors $\boldsymbol{\lambda}$ and $\boldsymbol{\nu}$, respectively. The resulting expression for the curvilineal divergence using λ components is:

$$\nabla^{\lambda}\bullet\mathbf{f}^{\lambda} = \nabla^{s}\bullet\mathbf{f}^{s} - \nabla^{s}\bullet(\boldsymbol{\nu}\mathbf{f}^{s})\bullet\boldsymbol{\nu} \tag{2.50a}$$

while the curvilineal divergence using the ν components is:

$$\nabla^{\nu}\bullet\mathbf{f}^{\nu} = \nabla^{s}\bullet\mathbf{f}^{s} - \nabla^{s}\bullet(\boldsymbol{\lambda}\mathbf{f}^{s})\bullet\boldsymbol{\lambda} \tag{2.50b}$$

Additionally, equations (2.47a) and (2.41) may be combined to eliminate $\nabla\bullet\mathbf{f}$ and obtain:

$$\nabla^{s}\bullet\mathbf{f}^{s} = \nabla^{\lambda}\bullet\mathbf{f}^{\lambda} + \nabla^{\nu}\bullet\mathbf{f}^{\nu} + \mathbf{n}\bullet\nabla\mathbf{n}\bullet\mathbf{f}^{s} + \boldsymbol{\lambda}\bullet\mathbf{f}^{s}\nabla\bullet\boldsymbol{\lambda} + \boldsymbol{\nu}\bullet\mathbf{f}^{s}\nabla\bullet\boldsymbol{\nu} \tag{2.50c}$$

Examples of the forms of the various terms appearing in the equations of this subsection appear in Table 2.2 for cartesian, cylindrical, and spherical systems.

2.3.8 Curl of a Spatial Vector

As with manipulations involving the divergence operator, the expression for the curl of a spatial vector $\mathbf{f}$ may be obtained with the orthogonal unit vectors $\boldsymbol{\lambda}$, $\mathbf{n}$, and $\boldsymbol{\nu}$ appearing explicitly. The expression for $\nabla\times\mathbf{f}$ is based on the definitions of ∇ and $\mathbf{f}$ given in equations (2.25) and (2.18), respectively, that may be invoked to obtain:

$$\nabla\times\mathbf{f} = (\boldsymbol{\lambda\lambda}\bullet\nabla + \mathbf{nn}\bullet\nabla + \boldsymbol{\nu\nu}\bullet\nabla)\times(\boldsymbol{\lambda\lambda}\bullet\mathbf{f} + \mathbf{nn}\bullet\mathbf{f} + \boldsymbol{\nu\nu}\bullet\mathbf{f}) \tag{2.51}$$

Expansion of this expression yields nine cross product terms that may be manipulated individually. This is readily accomplished using the identities involving the unit vectors listed in equations (2.8) through (2.14c). The result of this work is the following expression for the cross product:

$$\begin{aligned}\nabla\times\mathbf{f} = {} & \boldsymbol{\lambda}\,[\mathbf{n}\bullet\nabla(\boldsymbol{\nu}\bullet\mathbf{f}) - \boldsymbol{\nu}\bullet\nabla(\mathbf{n}\bullet\mathbf{f}) + \boldsymbol{\nu}\bullet\nabla\mathbf{n}\bullet\boldsymbol{\nu\nu}\bullet\mathbf{f} - \mathbf{n}\bullet\nabla\boldsymbol{\nu}\bullet\mathbf{nn}\bullet\mathbf{f}] \\ & + \mathbf{n}\,[\boldsymbol{\nu}\bullet\nabla(\boldsymbol{\lambda}\bullet\mathbf{f}) - \boldsymbol{\lambda}\bullet\nabla(\boldsymbol{\nu}\bullet\mathbf{f}) + \boldsymbol{\lambda}\bullet\nabla\boldsymbol{\nu}\bullet\boldsymbol{\lambda\lambda}\bullet\mathbf{f} - \boldsymbol{\nu}\bullet\nabla\boldsymbol{\lambda}\bullet\boldsymbol{\nu\nu}\bullet\mathbf{f}] \\ & + \boldsymbol{\nu}\,[\boldsymbol{\lambda}\bullet\nabla(\mathbf{n}\bullet\mathbf{f}) - \mathbf{n}\bullet\nabla(\boldsymbol{\lambda}\bullet\mathbf{f}) + \mathbf{n}\bullet\nabla\boldsymbol{\lambda}\bullet\mathbf{nn}\bullet\mathbf{f} - \boldsymbol{\lambda}\bullet\nabla\mathbf{n}\bullet\boldsymbol{\lambda\lambda}\bullet\mathbf{f}]\end{aligned} \tag{2.52}$$

This expansion is not as compact as the analogous expression for the divergence given by equation (2.38) primarily because the curl involves derivatives of components normal to the direction of the derivative. Nevertheless, the first two terms in each of the three groups of brackets are readily evaluated for a

Table 2.2: Geometric terms for divergence evaluation in standard coordinate systems

Coordinate System	$\boldsymbol{\lambda}$	$\mathbf{n}$	$\boldsymbol{\nu}$	$\nabla^{\lambda}\cdot\mathbf{f}^{\lambda}$	$\nabla^{n}\cdot\mathbf{f}^{n}$	$\nabla^{\nu}\cdot\mathbf{f}^{\nu}$	$\nabla\cdot\boldsymbol{\lambda}$	$\nabla\cdot\mathbf{n}$	$\nabla\cdot\boldsymbol{\nu}$	$\boldsymbol{\lambda}\cdot\nabla\boldsymbol{\lambda}$	$\mathbf{n}\cdot\nabla\mathbf{n}$	$\boldsymbol{\nu}\cdot\nabla\boldsymbol{\nu}$
Cartesian	$\mathbf{i}$	$\mathbf{j}$	$\mathbf{k}$	$\frac{\partial f_x}{\partial x}$	$\frac{\partial f_y}{\partial y}$	$\frac{\partial f_z}{\partial z}$	0	0	0	0	0	0
Cylindrical	$\mathbf{e}_r$	$\mathbf{e}_\theta$	$\mathbf{e}_z$	$\frac{\partial f_r}{\partial r}$	$\frac{1}{r}\frac{\partial f_\theta}{\partial \theta}$	$\frac{\partial f_z}{\partial z}$	$\frac{1}{r}$	0	0	0	$-\frac{\mathbf{e}_r}{r}$	0
Spherical	$\mathbf{e}_r$	$\mathbf{e}_\theta$	$\mathbf{e}_\phi$	$\frac{\partial f_r}{\partial r}$	$\frac{1}{r}\frac{\partial f_\theta}{\partial \theta}$	$\frac{1}{r\sin\theta}\frac{\partial f_\phi}{\partial \phi}$	$\frac{2}{r}$	$\frac{1}{r\tan\theta}$	0	0	$-\frac{\mathbf{e}_r}{r}$	$-\frac{\mathbf{e}_r}{r}-\frac{\mathbf{e}_\theta}{r\tan\theta}$

general orthogonal coordinate system while the last two terms in each set of brackets account for the curvature of the coordinates. As examples, expressions for the derivative operators and for the derivatives of the unit vectors for cartesian, cylindrical, and spherical coordinates in equation (2.52) are given in Table 2.3.

2.3.9 Surface Curl of a Surface Vector

For a surface with unit normal vector $\mathbf{n}$, the surface curl of a vector $\mathbf{f}^s = \mathbf{f} - \mathbf{nn} \bullet \mathbf{f}$ tangent to the surface is related to the spatial curl of $\mathbf{f}$ by:

$$\nabla^s \times \mathbf{f}^s = (\nabla - \mathbf{nn} \bullet \nabla) \times (\mathbf{f} - \mathbf{nn} \bullet \mathbf{f}) \tag{2.53}$$

Expansion of the right side of this expression using equation (2.11) which provides $\mathbf{n} \times \mathbf{n} = 0$ yields:

$$\nabla^s \times \mathbf{f}^s = \nabla \times \mathbf{f} + \mathbf{n} \bullet \nabla \mathbf{f} \times \mathbf{n} - (\nabla \times \mathbf{n})\, \mathbf{n} \bullet \mathbf{f} + \mathbf{n} \times \nabla (\mathbf{n} \bullet \mathbf{f}) - \mathbf{n} \bullet \nabla \mathbf{n} \times \mathbf{nn} \bullet \mathbf{f} \tag{2.54}$$

By equation (2.15b), the third and fifth terms cancel so that the final relation obtained is:

$$\nabla^s \times \mathbf{f}^s = \nabla \times \mathbf{f} + \mathbf{n} \bullet \nabla \mathbf{f} \times \mathbf{n} + \mathbf{n} \times \nabla (\mathbf{n} \bullet \mathbf{f}) \tag{2.55}$$

2.3.10 Curvilineal Curl of Curvilineal Vector

For a curve with unit tangent vector $\boldsymbol{\lambda}$, the curvilineal curl, ∇^c, of a vector $\mathbf{f}^c = \boldsymbol{\lambda\lambda} \bullet \mathbf{f}$ tangent to the curve may be related to the spatial curl of $\mathbf{f}$. Equations (2.20) and (2.27) may be invoked to obtain:

$$\nabla^c \times \mathbf{f}^c = (\boldsymbol{\lambda\lambda} \bullet \nabla) \times (\boldsymbol{\lambda\lambda} \bullet \mathbf{f}) \tag{2.56}$$

Because by equation (2.11), $\boldsymbol{\lambda} \times \boldsymbol{\lambda} = 0$, this equation reduces to:

$$\nabla^c \times \mathbf{f}^c = -(\boldsymbol{\lambda} \bullet \nabla \boldsymbol{\lambda}) \times \boldsymbol{\lambda\lambda} \bullet \mathbf{f} \tag{2.57}$$

Note that if the curve is a straight line, $\boldsymbol{\lambda} \bullet \nabla \boldsymbol{\lambda} = 0$ and $\nabla^c \times \mathbf{f}^c$ will equal 0.

2.3.11 Miscellaneous Relationships Involving the Curl

Manipulations with the curl operator tend to be tedious and thus it will be convenient to have available some relations among the spatial, surficial, and curvilineal curl in addition to those in the preceding sections. The spatial curl of a spatial quantity may be expressed in terms of the components as:

Table 2.3: Geometric terms for curl evaluation in standard coordinate systems

Coordinate System	$\boldsymbol{\lambda}$	$\mathbf{n}$	$\boldsymbol{\nu}$	$\boldsymbol{\lambda}\bullet\nabla$	$\mathbf{n}\bullet\nabla$	$\boldsymbol{\nu}\bullet\nabla$	$\mathbf{n}\bullet\nabla\boldsymbol{\lambda}$	$\boldsymbol{\nu}\bullet\nabla\boldsymbol{\lambda}$	$\boldsymbol{\lambda}\bullet\nabla\mathbf{n}$	$\boldsymbol{\nu}\bullet\nabla\mathbf{n}$	$\boldsymbol{\lambda}\bullet\nabla\boldsymbol{\nu}$	$\mathbf{n}\bullet\nabla\boldsymbol{\nu}$
Cartesian	$\mathbf{i}$	$\mathbf{j}$	$\mathbf{k}$	$\frac{\partial}{\partial x}$	$\frac{\partial}{\partial y}$	$\frac{\partial}{\partial z}$	0	0	0	0	0	0
Cylindrical	$\mathbf{e}_r$	$\mathbf{e}_\theta$	$\mathbf{e}_z$	$\frac{\partial}{\partial r}$	$\frac{1}{r}\frac{\partial}{\partial \theta}$	$\frac{\partial}{\partial z}$	$\frac{\mathbf{e}_\theta}{r}$	0	0	0	0	0
Spherical	$\mathbf{e}_r$	$\mathbf{e}_\theta$	$\mathbf{e}_\phi$	$\frac{\partial}{\partial r}$	$\frac{1}{r}\frac{\partial}{\partial \theta}$	$\frac{1}{r\sin\theta}\frac{\partial}{\partial \phi}$	$\frac{\mathbf{e}_\theta}{r}$	$\frac{\mathbf{e}_\phi}{r}$	0	$\frac{\mathbf{e}_\phi}{r\tan\theta}$	0	0

$$\nabla\times\mathbf{f} = (\nabla^{\lambda}+\nabla^{n}+\nabla^{\nu})\times(\mathbf{f}^{\lambda}+\mathbf{f}^{n}+\mathbf{f}^{\nu}) \tag{2.58}$$

where, for example, $\nabla^{\lambda}\equiv\boldsymbol{\lambda}\boldsymbol{\lambda}\boldsymbol{\cdot}\nabla$. Expansion of this expression and use of equations analogous to equation (2.57) for the case when the del operator and the component vector are in the same direction yields:

$$\begin{aligned}\nabla\times\mathbf{f} = {} & [\,(\nabla^{\lambda}\times\mathbf{f}^{n}) + (\nabla^{n}\times\mathbf{f}^{\lambda})\,] + [\,(\nabla^{n}\times\mathbf{f}^{\nu}) + (\nabla^{\nu}\times\mathbf{f}^{n})\,] \\ & + [\,(\nabla^{\nu}\times\mathbf{f}^{\lambda}) + (\nabla^{\lambda}\times\mathbf{f}^{\nu})\,] - \boldsymbol{\lambda}\boldsymbol{\cdot}\nabla\boldsymbol{\lambda}\times\mathbf{f}^{\lambda} - \mathbf{n}\boldsymbol{\cdot}\nabla\mathbf{n}\times\mathbf{f}^{n} - \boldsymbol{\nu}\boldsymbol{\cdot}\nabla\boldsymbol{\nu}\times\mathbf{f}^{\nu}\end{aligned} \tag{2.59}$$

For the case of cartesian coordinates, the last three terms in this equation will be zero and the three bracketed terms account for the $\boldsymbol{\nu}$, $\boldsymbol{\lambda}$, and $\mathbf{n}$ components of the curl, respectively.

A second relation that is useful is an expansion of the surficial curl of a surface vector that is tangent to the surface. If the normal to the surface is $\mathbf{n}$, the expression under consideration is:

$$\nabla^{s}\times\mathbf{f}^{s} = (\nabla^{\nu}+\nabla^{\lambda})\times(\mathbf{f}^{\nu}+\mathbf{f}^{\lambda}) \tag{2.60}$$

Manipulation of the expressions on the right side of this equation making use of equations (2.8) through (2.15c) yields:

$$\nabla^{s}\times\mathbf{f}^{s} = [\,(\boldsymbol{\lambda}\boldsymbol{\nu}-\boldsymbol{\nu}\boldsymbol{\lambda})\boldsymbol{\cdot}\nabla^{s}\mathbf{n}\boldsymbol{\cdot}\mathbf{f}^{s}] + [\mathbf{n}\,(\boldsymbol{\nu}\boldsymbol{\cdot}\nabla^{s}\mathbf{f}^{s}\boldsymbol{\cdot}\boldsymbol{\lambda} - \boldsymbol{\lambda}\boldsymbol{\cdot}\nabla^{s}\mathbf{f}^{s}\boldsymbol{\cdot}\boldsymbol{\nu})\,] \tag{2.61a}$$

The first term in brackets accounts for curvature of the coordinates while the second term is the component of the curl of the surface vector in the direction normal to the surface. Equation (2.61a) may be written in another form by using the relation $\mathbf{n}\times\mathbf{I} = \boldsymbol{\lambda}\boldsymbol{\nu}-\boldsymbol{\nu}\boldsymbol{\lambda}$ where $\mathbf{I}$ is the identity tensor to obtain:

$$\nabla^{s}\times\mathbf{f}^{s} = [\,(\mathbf{n}\times\mathbf{I})\boldsymbol{\cdot}\nabla^{s}\mathbf{n}\boldsymbol{\cdot}\mathbf{f}^{s}] - [\mathbf{n}\,(\mathbf{n}\times\mathbf{I}):\nabla^{s}\mathbf{f}^{s}] \tag{2.61b}$$

The fact that only the vector in the normal direction appears explicitly in this last equation is expected since the relation must hold for any orthogonal $\lambda-\nu$ coordinates selected to lie in the surface.

2.4 TIME DERIVATIVES

Integration over various domains of derivatives of a function changes the scale of these derivatives. Many of the integration theorems relate integrals of partial time derivatives to time derivatives of integrals over volumes, surfaces, and curves. These partial derivatives are taken fixed to a position in space, on a surface, or along a curve. When the time derivative is moved outside the integral, the differentiation is such that it applies to the integration region, a region that may be moving through space. This feature must be accounted for in the development of relations among various time derivatives.

2.4.1 Partial Time Derivative in Space

The partial derivative with respect to time of a spatial function, $f(\mathbf{x}, t)$, that is dependent on three spatial coordinates and time is written:

$$\frac{\partial f(\mathbf{x}, t)}{\partial t} = \left.\frac{\partial f}{\partial t}\right|_{\mathbf{x}} \tag{2.62}$$

where the vertical bar with subscript $\mathbf{x}$ indicates that the spatial coordinates are held constant. The partial derivative in equation (2.62) measures the change in a function with time at a point fixed in space.

If a time derivative of a function is to be calculated while moving through space at some velocity $d\mathbf{x}/dt = \mathbf{w}$, this rate of change of coordinate position must be accounted for. Rather than being a partial time derivative fixed in space, this is a total time derivative, df/dt, and is calculated as [*Whitaker*, 1968]:

$$\frac{df}{dt} = \left.\frac{\partial f}{\partial t}\right|_{\mathbf{x}} + \mathbf{w} \bullet \nabla f \tag{2.63}$$

The partial time derivative of f is local rate of change of f, while the last term accounts for changes observed in f due to the change with time of the location at which f is measured.

2.4.2 Partial Time Derivative on a Surface

A function, f, may be defined that is a property of a surface. Furthermore, it may be desirable to calculate the time rate of change of that function as the surface moves and deforms in space. A function constrained to be evaluated at a position on a surface may be defined with time, t, and two surficial coordinates, $\mathbf{u}$, as independent variables such that $f = f(\mathbf{u}, t)$ (Note that the surficial coordinates may be expressed in terms of spatial coordinates such that $\mathbf{u} = \mathbf{u}(\mathbf{x})$). The time derivative of this function at a position fixed on the surface (i.e. with $\mathbf{u}$ coordinates held constant) is written:

$$\frac{\partial f(\mathbf{u}, t)}{\partial t} = \left.\frac{\partial f}{\partial t}\right|_{\mathbf{u}} \tag{2.64}$$

where the vertical bar with subscript $\mathbf{u}$ indicates that the surficial coordinates are held constant. Note that the partial derivative in equation (2.64) mandates that, although the surface coordinates are being held constant, movement in the direction normal to the surface at the velocity of the surface would be necessary in order to remain on the surface and measure the indicated partial derivative.

If the time derivative of $f(\mathbf{u}, t)$ is to be calculated while moving in the surface at some velocity $d\mathbf{u}/dt = \mathbf{w}^s$, the rate of change of surficial coordinate position must be accounted for. This time derivative of f is also a total time derivative and may be expressed as:

$$\frac{df}{dt} = \left.\frac{\partial f}{\partial t}\right|_{\mathbf{u}} + \mathbf{w}^s \bullet \nabla^s f \tag{2.65}$$

where ∇^s is the surficial gradient operator given by equation (2.26) when $\mathbf{n}$ is the unit normal to the surface.

Note that the derivatives on the left side of equations (2.63) and (2.65) are both total derivatives. If $\mathbf{nn} \bullet \mathbf{w}$ is the normal velocity of the surface such that $\mathbf{w} = \mathbf{w}^s + \mathbf{nn} \bullet \mathbf{w}$, these total derivatives will be equal and:

$$\left.\frac{\partial f}{\partial t}\right|_{\mathbf{x}} + \mathbf{w} \bullet \nabla f = \left.\frac{\partial f}{\partial t}\right|_{\mathbf{u}} + \mathbf{w}^s \bullet \nabla^s f \tag{2.66}$$

Because $\mathbf{w}^s$ has no components in the $\mathbf{n}$ direction, the following equalities apply:

$$\mathbf{w}^s \bullet \nabla^s = \mathbf{w}^s \bullet \nabla = \mathbf{w} \bullet \nabla - \mathbf{w} \bullet \mathbf{nn} \bullet \nabla \tag{2.67}$$

and equation (2.66) simplifies to:

$$\left.\frac{\partial f}{\partial t}\right|_{\mathbf{x}} = \left.\frac{\partial f}{\partial t}\right|_{\mathbf{u}} - \mathbf{w} \bullet \mathbf{nn} \bullet \nabla f \tag{2.68}$$

This relation between the time derivative of a function fixed in space to the time derivative of a function fixed to a point on a surface will prove useful in subsequent theorem derivation. Note that for the case where the normal velocity of the surface is zero, the two partial time derivatives are equal.

2.4.3 Partial Time Derivative on a Curve

A function f may be defined in terms of a position along a simple curve and time. If the lineal coordinate along the curve is denoted as the $\mathbf{l}$ coordinate, the function may be written as $f(\mathbf{l}, t)$. The single curvilineal coordinate, $\mathbf{l}$, indicating position along the curve may be expressed in terms of spatial coordinates such that $\mathbf{l} = \mathbf{l}(\mathbf{x})$ or in terms of surficial coordinates when the curve lies in a surface such that $\mathbf{l} = \mathbf{l}(\mathbf{u})$. The partial time derivative, corresponding to the time derivative at a fixed curvilineal position on the curve is written:

$$\frac{\partial f(\mathbf{l}, t)}{\partial t} = \left.\frac{\partial f}{\partial t}\right|_{\mathbf{l}} \tag{2.69}$$

For this time derivative, although the coordinate along the curve is held constant, the evaluation takes place while moving with the curve as it translates in directions normal to $\mathbf{l}$.

The total time derivative is the rate of change of f calculated while moving at some velocity $d\mathbf{l}/dt = \mathbf{w}^c$ along the curve. Note that $\mathbf{w}^c$ is tangent to the curve and may also be written as $\mathbf{w}^c = \boldsymbol{\lambda\lambda}\bullet\mathbf{w}$ when $\boldsymbol{\lambda}$ is the unit vector tangent to the curve. Thus the total time derivative is:

$$\frac{df}{dt} = \left.\frac{\partial f}{\partial t}\right|_{\mathbf{l}} + \mathbf{w}^c\bullet\nabla^c f \tag{2.70}$$

where ∇^c is the gradient operator along the curve given by equation (2.27). If the velocity at which the curve moves through space in a direction normal to the curve is indicated as $\mathbf{w} - \mathbf{w}^c$, the total derivatives of f given by equations (2.63) and (2.70) will be equal such that:

$$\left.\frac{\partial f}{\partial t}\right|_{\mathbf{x}} + \mathbf{w}\bullet\nabla f = \left.\frac{\partial f}{\partial t}\right|_{\mathbf{l}} + \mathbf{w}^c\bullet\nabla^c f \tag{2.71}$$

Because $\mathbf{w}^c\bullet\nabla^c = \mathbf{w}^c\bullet\nabla = \mathbf{w}\bullet\boldsymbol{\lambda\lambda}\bullet\nabla$, this equation may be rearranged to the form:

$$\left.\frac{\partial f}{\partial t}\right|_{\mathbf{x}} = \left.\frac{\partial f}{\partial t}\right|_{\mathbf{l}} - (\mathbf{w} - \mathbf{w}\bullet\boldsymbol{\lambda\lambda})\bullet\nabla f \tag{2.72}$$

This equation expresses the relationship between a time derivative of f at a point fixed in space to the time derivative of f at a point fixed on a curve that moves through space with velocity $\mathbf{w} - \mathbf{w}\bullet\boldsymbol{\lambda\lambda}$.

By equating the total time derivatives in equations (2.65) and (2.70) and employing similar manipulations to those above, one may obtain the following relations between the time derivative of a function at a point on a curve and the time derivative of the function at a point on the surface containing the curve:

$$\left.\frac{\partial f}{\partial t}\right|_{\mathbf{u}} = \left.\frac{\partial f}{\partial t}\right|_{\mathbf{l}} - (\mathbf{w}^s - \mathbf{w}^s\bullet\boldsymbol{\lambda\lambda})\bullet\nabla^s f \tag{2.73a}$$

$$\left.\frac{\partial f}{\partial t}\right|_{\mathbf{u}} = \left.\frac{\partial f}{\partial t}\right|_{\mathbf{l}} - (\mathbf{w}^s - \mathbf{w}^s\bullet\boldsymbol{\lambda\lambda})\bullet\nabla f \tag{2.73b}$$

$$\left.\frac{\partial f}{\partial t}\right|_{\mathbf{u}} = \left.\frac{\partial f}{\partial t}\right|_{\mathbf{l}} - (\mathbf{w} - \mathbf{w}\bullet\boldsymbol{\lambda\lambda})\bullet\nabla^s f \tag{2.73c}$$

where $\mathbf{w}^s$ is the velocity of the surface and ∇^s is the surface gradient operator. The alternative equations (2.73b) and (2.73c) arise because the dot product between the velocity and the gradient operator is unchanged by adding a component normal to the surface to either of these vectors.

2.5 CONCLUSION

Within this chapter the mathematical preliminaries essential for development of the integration and averaging and transport theorems have been set forth. A prime goal of the presentation was to show relationships between various types of spatial derivatives and between various types of time derivatives. A general set of orthogonal unit vectors in space was employed in developing expressions in vector operator notation for the gradient, divergence, and curl of functions in curvilinear coordinates. The discussion included spatial, surficial, and curvilineal operators. Finally, the time derivative of a function was presented. The partial derivative of a function has different meanings depending on whether it is calculated holding spatial, surficial, or curvilineal coordinates constant. The total derivative was employed in conjunction with these three representations to obtain relations among the various partial derivatives.

Command and understanding of the mathematical preliminaries contained in this chapter are essential to the manipulations in Chapter 3 dealing with generalized functions for space, surfaces, and curves and to the presentation in Chapter 4 of the various scales considered in the derivation of the integral theorems. These preliminary concepts are also employed in the example theorem derivations of Chapter 6 and are needed to derive the theorems tabulated in Chapters 7 and 8. Finally, in the applications of Chapter 9, the mathematical framework presented here is used in deriving balance equations for specific physical problems.

2.6 REFERENCES

Aris, R., *Vectors, Tensors, and the Basic Equations of Fluid Mechanics*, Prentice-Hall, Englewood Cliffs, 1962.

Gray, W. G., and S. M. Hassanizadeh, Averaging theorems and averaged equations for transport of interface properties in multiphase systems, *International Journal of Multiphase Flow*, **15**(1), 81-95, 1989.

Hildebrand, F. B., *Advanced Calculus for Applications*, Second edition, Prentice-Hall, Englewood Cliffs, 1976.

Kreyszig, E., *Advanced Engineering Mathematics*, Wiley, New York, 1962.

Moeckel, G. P., Thermodynamics of an interface, *Archive for Rational Mechanics and Analysis*, **57**, 255-280, 1975.

Whitaker, S., *Introduction to Fluid Mechanics*, Prentice-Hall, Englewood Cliffs, 1968.

CHAPTER THREE

BASIC CONCEPTS FOR GENERALIZED FUNCTIONS

3.0 INTRODUCTION

One of the prime objectives of this work is the development of two classes of theorems: 1) integration theorems and 2) averaging theorems. In the derivation of these theorems, it is often required to interchange the order of integration and differentiation. Also, a change of scale accompanies this interchange, such as transformation of a three-dimensional microscopic problem to a three-dimensional macroscopic problem. The manipulations required to produce the relations between derivatives at different scales and between integrals of derivatives and derivatives of integrals are most easily performed in the context of generalized functions.

Generalized functions provide a simple method to move spatial and temporal dependence of the limits of integration into the integrand. Also, the integral properties of generalized functions provide a simplified route to changes in scale. Once generalized functions are introduced into a theorem derivation, all subsequent manipulations involve the integrand with minimal concern for the limits of integration. This can be contrasted to more conventional methods for derivation of some of the theorems of interest [e.g. *Whitaker*, 1968; *Slattery*, 1972] which keep track of moving, deforming regions and determine changes to the system as the region moves through space. For a single phase system, this latter technique is tedious at best; for multiphase systems with interfaces and contact lines, the conventional approach becomes unwieldy. In the current context, the power of generalized functions lies in their streamlining of the derivation of integration and averaging theorems for volumes, surfaces, and curves.

A rigorous mathematical development of the theory of generalized functions is beyond the scope of this work. Ample references exist for this purpose [*Schwartz*, 1950, 1951; *Lighthill*, 1958; *Gel'fand and Shilov*, 1964; *Kanwal*, 1983]. Instead, this chapter focuses on properties of generalized functions that are useful in deriving integration and averaging theorems. Hence, many of the properties of generalized functions will be presented from a descriptive point of view. First, basic concepts will be illustrated using the simplest scenarios, the one-dimensional, straight line setting. Next, these ideas will be extended to general space curves, surfaces, and arbitrary three-dimensional volumes. For each of these geometries, basic definitions for the generalized function will be given, space and time derivative properties will be derived, and integral properties will be established. In Chapter 4 of the text, examples using generalized functions to delineate regions of interest will be provided. In Chapter 6, some important integral identities needed to change scales of a problem will be set forth.

3.1 GENERALIZED FUNCTIONS IN ONE DIMENSION: STRAIGHT LINE CASE

The basic motivation for the use of generalized functions to develop integration and averaging theorems for the gradient, divergence, curl, and time derivative operator is quite simple. These functions can be used to identify a region of interest by taking on the value of one in the interior and zero exterior to the region. The value of the generalized function at the boundary of the region, where it undergoes a step change in value from one to zero, is unimportant as long as it is finite. The generalized function can be used to designate the domain of any function as the region of interest by simply forming the product of the generalized function with the function under consideration.

For concreteness, this section will focus on the use of generalized functions to select portions of a straight line in one coordinate direction. In this setting, the generalized function to be employed is an extension of the unit step function, sometimes called the Heaviside step function. Although the defining sequence of functions for the Heaviside step function, $U(x-x_0)$, is not unique, one suitable definition is given as follows [*Lighthill*, 1958]:

$$U(x-x_0) = \lim_{m \to \infty} \{g_m\} \quad \textbf{(3.1a)}$$

where:

$$g_m = \frac{1}{2} + \frac{1}{2}\operatorname{erf}\left[\sqrt{m}(x-x_0)\right] \quad \textbf{(3.1b)}$$

and:

$$\text{erf}\,(x) = \frac{2}{\sqrt{\pi}}\int_0^x e^{-u^2}du \qquad \textbf{(3.1c)}$$

The Heaviside function can also depend on time by allowing $x_0 = x_0(t)$.

A plot of g_m for various values of m is shown in figure 3.1. The sequence converges to the unit step function. In the limit $m \to \infty$, it can be shown that $\text{erf}\,[\sqrt{m}\,(x-x_0)] \to 1$ for $x > x_0$ and $\text{erf}\,[\sqrt{m}\,(x-x_0)] \to -1$ for $x < x_0$ so that $U(x-x_0)$ does equal g_∞. Although this sequence is not uniformly convergent, it is pointwise convergent, a property which allows it to be applied subsequently in the development of the theorems of this book.

Now a generalized function, which will be denoted γ^L for use with a straight line segment, can be defined in terms of the step function such that it identifies a portion of a straight line in the domain $a < x < b$ as:

$$\gamma^L(x, t) = U(x-a) - U(x-b) \qquad \textbf{(3.2)}$$

where the dependence of γ^L on time would occur if either a or b were a function of time. Note that by this definition, γ^L is equal to 1 for $a < x < b$ and 0

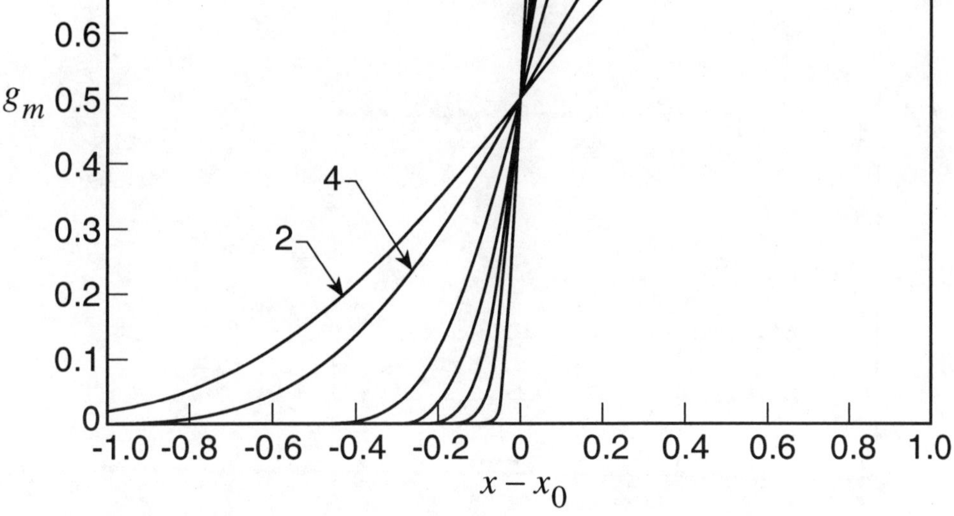

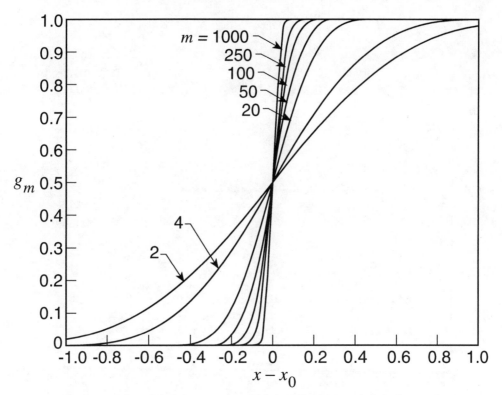

Figure 3.1. Plot of defining sequence for the Heaviside step function, $g_m(x\text{-}x_0)$, as given in equation (3.1b).

outside of this domain on the line as shown in figure 3.2 (In the electrical engineering literature, this function is commonly referred to as a gate function when the independent variable is time.). Note that $\gamma^L(x, t)$ in equation (3.2) can be extended to indicate any number of non-overlapping domains along the x-axis without changing any of its properties.

To use the generalized function to isolate the behavior of a function to the particular region of interest, simply form the product $\gamma^L f(x, t)$ where $f(x, t)$ is the function of interest. This new function has the properties:

$$\gamma^L f(x, t) = \begin{cases} 0 & x < a \\ f(x, t) & a < x < b \\ 0 & x > b \end{cases} \tag{3.3}$$

Inclusion of the endpoints is irrelevant since all manipulations of γ^L will be carried out under integration and the value of an integral is not affected by a finite value of the integrand at a point. Based on equation (3.3), the following equivalence between integrals may be stated:

$$\int_a^b f(x, t)\, dx = \int_{-\infty}^{\infty} \gamma^L f(x, t)\, dx \tag{3.4}$$

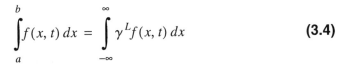

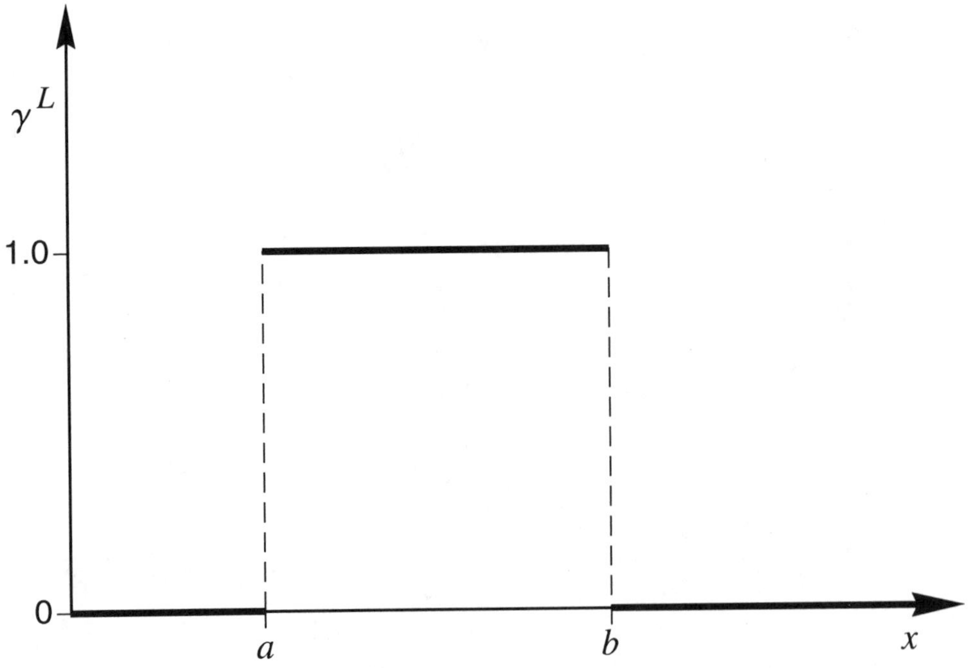

Figure 3.2. Plot of function γ^L vs. x indicating the region $a<x<b$ as the one-dimensional region of interest along the x-axis.

The change of the limits of integration from a finite region of space to an infinite domain will prove particularly useful in subsequent derivation of theorems. Note that the possible temporal dependence of the integration limits on the left side of equation (3.4) has been shifted to the integrand on the right side of this expression where the integration limits are independent of time.

3.1.1 Spatial Derivative of the One-Dimensional Generalized Function

When generalized functions are used to develop integration and averaging theorems, the expressions $\partial\gamma^L/\partial x$ and $\partial\gamma^L/\partial t$ arise. As a preliminary step to the study of these derivatives, differentiate the Heaviside step function, as given in equation (3.1a), with respect to x and interchange the order of differentiation and the limit process to obtain:

$$\frac{\partial U}{\partial x} = \lim_{m\to\infty}\frac{\partial g_m}{\partial x} = \lim_{m\to\infty}\left\{\sqrt{\frac{m}{\pi}}\exp\left[-m(x-x_0)^2\right]\right\} \qquad \textbf{(3.5a)}$$

This expression defines the generalized function known as the Dirac delta function [*Lighthill*, 1958], and the sequence is depicted in figure 3.3. Other sequences may alternatively be used that, in some limit, define the Dirac delta function [see, e.g., *Kanwal*, 1983]. For the one-dimensional straight line case, the delta function is symbolized by:

$$\frac{\partial U}{\partial x} = \delta(x-x_0) \qquad \textbf{(3.5b)}$$

where x_0 may be a function of time. Application of this relation to the derivative of γ^L, as defined in equation (3.2) where $a<x<b$ is the region of interest, yields:

$$\frac{\partial\gamma^L}{\partial x} = \delta(x-a) - \delta(x-b) \qquad \textbf{(3.6)}$$

Because the sequence of functions defined by g_m in equation (3.1b) does not converge uniformly, interchanging the limit and derivative operations requires an extension of standard mathematical theory. *Dirac* [1930] recognized the utility of the delta function so he carried out the differentiation process cognizant that the ensuing result no longer falls under the guise of conventional mathematics. Subsequently, *Schwartz* [1950; 1951] introduced the Theory of Distributions which provides rigorous theoretical support for Dirac's work.

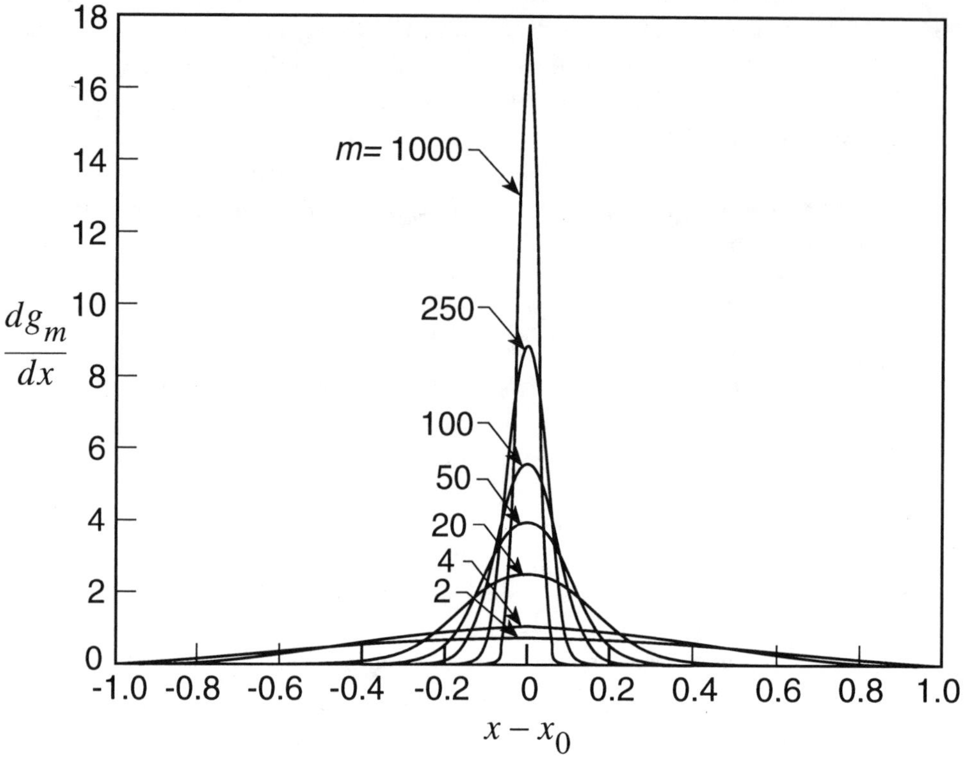

Figure 3.3. Plot of defining sequence for the Dirac delta function, $dg_m(x\text{-}x_0)/dx$, as given in equation (3.5a).

The Dirac delta function and its extensions are accepted and used in many fields of science and mathematics. For example, in hydrology, the delta function is used to indicate point sources and sinks of water. In electrical engineering, it is used in describing a unit impulse applied to a circuit. In structural engineering, the delta function is employed in representing point loadings on structural members while in physics, it is used when representing point charges. In applied mathematics, the Dirac function is used as a tool in Fourier analysis and in solving elliptic partial differential equations using Green's functions. *Gray and Celia* [1990] have demonstrated the use of the generalized function in a variety of engineering settings, including the application of the finite element method. Common use of the function is not a rigorous proof of validity, but as *Hildebrand* [1976] states concerning their utility, "they are not true functions, … nevertheless if formal use of them leads to a result which is capable of physical interpretation, then in practical cases the result may be accepted as correct." The present work will carry on in this spirit.

3.1.2 Integral Properties of $\partial\gamma^L/\partial x$

The delta function, $\delta(x - x_0)$, is non-zero only at $x = x_0$. Its behavior at $x = x_0$ is required to be such that:

$$\int_{-\infty}^{\infty} \delta(x - x_0)\, dx \equiv 1 \tag{3.7}$$

In some presentations, the delta function is considered to "equal" infinity at $x = x_0$ although *Kanwal* [1983] shows that this is not a necessary constraint. In the present case, the value of this function at x_0 is considered to be unique but unspecified. However, the integral properties in the vicinity of x_0 are important. From equation (3.7), it can be shown, using equation (3.5b) and integration by parts, that:

$$\int_{-\infty}^{\infty} f(x, t)\, \delta(x - x_0)\, dx = f(x_0, t) \tag{3.8}$$

Thus, with $\partial\gamma^L/\partial x$ defined as in equation (3.6):

$$-\int_{-\infty}^{\infty} f(x, t) \frac{\partial\gamma^L}{\partial x} dx = f(b, t) - f(a, t) \tag{3.9}$$

3.1.3 Time Derivative of γ^L

For proving theorems, the time derivative of γ^L at a point fixed on the line, $\partial\gamma^L/\partial t|_x$, must be expressed in terms of the spatial derivative. Consider the total derivative of γ^L with respect to time. By definition:

$$\frac{d\gamma^L}{dt} = \left.\frac{\partial\gamma^L}{\partial t}\right|_x + \frac{\partial\gamma^L}{\partial x}\frac{dx}{dt} \tag{3.10}$$

where dx/dt is the rate of change of position (i.e the velocity of an observer) as $d\gamma^L/dt$ is evaluated; and $\partial\gamma^L/\partial x$ is given by equation (3.6). In equation (3.10), the first term on the right hand side represents the rate of change of γ^L at a fixed point on the line, while the second term accounts for changes observed in γ^L as the point of evaluation moves along the line.

The quantity $d\gamma^L/dt$, therefore, represents the total change in γ^L that an observer moving with velocity dx/dt would see. Recall that γ^L is used to specify the region of interest $a < x < b$ where the boundary points of this region, a and b, may be functions of time. Further, note that dx/dt is the velocity of an observer at any point in the domain $-\infty < x < \infty$ and may exhibit spatial and/or temporal variations. For an arbitrary velocity, $d\gamma^L/dt$ will, in general, be non-zero. However, $d\gamma^L/dt$ will be zero if the velocity is constrained such that $dx/dt = w_x$ where w_x is defined as follows:

1. $w_x|_a = da/dt$ and $w_x|_b = db/dt$ (i.e. an observer at the boundary of the region of interest moves with the boundary as the region deforms); and
2. w_x is chosen such that an observer not on the boundary never crosses the boundary.

An observer constrained to move at these velocities will detect no change in the value of γ^L with time. For example, an observer located at $x = (a+b)/2$ moving under the above constraints always sees γ^L having a value of 1. Thus, with the above constraints enforced, $d\gamma^L/dt = 0$; and from equation (3.10), the following important relation is obtained:

$$\left.\frac{\partial\gamma^L}{\partial t}\right|_x = -\frac{\partial\gamma^L}{\partial x}w_x \tag{3.11}$$

This equation relates the time derivative of γ^L at a point to the spatial derivative defined previously in equation (3.6). Recall that $\partial\gamma^L/\partial x$ is nonzero only at the boundary of the region of interest. Thus identification of w_x as the velocity of the boundary is very important and useful in obtaining the theorems of interest here.

3.1.4 Example Using γ^L to Prove Leibnitz' Rule

Before extending the development of generalized functions to multiple spatial dimensions, an example making use of generalized functions to prove averaging theorems may prove enlightening. Specifically, a proof of Leibnitz' rule, a theorem relating the integral of a derivative to the derivative of the integral, will be presented.

Consider a function $f(x, t)$ where, for convenience, x is considered as the spatial coordinate and t is time. The objective of this example is to relate the integral of the derivative given by:

$$\int_{a(t)}^{b(t)}\left.\frac{\partial f}{\partial t}\right|_x dx \tag{3.12}$$

to the derivative of the integral. Even though the variable of integration is x and the differentiation is with respect to t, the order of differentiation and integration cannot be simply interchanged because the limits of integration depend on t. To transfer this dependence from the limits of integration to the integrand, use will be made of the generalized function defined by:

$$\gamma^L(x,t) = \begin{cases} 0 & x < a(t) \\ 1 & a(t) < x < b(t) \\ 0 & x > b(t) \end{cases} \qquad \textbf{(3.13)}$$

Thus, γ^L depends on t indirectly through a and b. Introduction of definition (3.13) into integral (3.12) allows the limits of integration to be extended to include the entire x domain such that:

$$\int_{a(t)}^{b(t)} \left.\frac{\partial f}{\partial t}\right|_x dx = \int_{-\infty}^{\infty} \left.\frac{\partial f}{\partial t}\right|_x \gamma^L dx \qquad \textbf{(3.14)}$$

Now the chain rule may be applied to the right side of this equation to obtain the equality:

$$\int_{a(t)}^{b(t)} \left.\frac{\partial f}{\partial t}\right|_x dx = \int_{-\infty}^{\infty} \left.\frac{\partial (f\gamma^L)}{\partial t}\right|_x dx - \int_{-\infty}^{\infty} \left. f\frac{\partial \gamma^L}{\partial t}\right|_x dx \qquad \textbf{(3.15)}$$

In the first term on the right side of equation (3.15), the order of differentiation and integration can be reversed directly because the limits of integration are now independent of time. Additionally, equation (3.11) can be applied to the second term on the right side to change the time derivative to a spatial derivative. Thus equation (3.15) becomes:

$$\int_{a(t)}^{b(t)} \left.\frac{\partial f}{\partial t}\right|_x dx = \frac{\partial}{\partial t}\int_{-\infty}^{\infty} f\gamma^L dx + \int_{-\infty}^{\infty} f w_x \frac{\partial \gamma^L}{\partial x} dx \qquad \textbf{(3.16)}$$

The first term on the right side of this equation will now be rearranged to its final form. First, it should be noted that explicit indication that differentiation with respect to time must be performed while holding x constant has been dropped because the integrated expression will not depend on x. Since the differentiation has been extracted from the integrand, the limits of integration may now be conveniently changed back to their original values. Because of the definition of γ^L provided in equation (3.13), the integrand is nonzero only for $a < x < b$, and is simply equal to f in this region. One subtle but very important point is the change of the partial derivative to a total derivative. In the current case, this change may seem trivial because the evaluated integral is a function of t only. However in a more general case, when the limits of integration are altered to reflect a change from integration over all space to integration over some finite domain, the use of d/dt rather than $\partial/\partial t$ indicates that the domain is being followed. With these considerations, equation (3.16) becomes:

$$\int_{a(t)}^{b(t)} \left.\frac{\partial f}{\partial t}\right|_x dx = \frac{d}{dt}\int_{a(t)}^{b(t)} f dx + \int_{-\infty}^{\infty} f w_x \frac{\partial \gamma^L}{\partial x} dx \tag{3.17}$$

The second integral on the right side of this equation may be evaluated directly by applying equation (3.9) to obtain:

$$\int_{a(t)}^{b(t)} \left.\frac{\partial f}{\partial t}\right|_x dx = \frac{d}{dt}\int_{a(t)}^{b(t)} f dx - (f w_x)\big|_{x=b} + (f w_x)\big|_{x=a} \tag{3.18}$$

The constraints preceding equation (3.11) indicate that:

$$w_x\big|_{x=b} = \frac{db}{dt} \tag{3.19a}$$

and:

$$w_x\big|_{x=a} = \frac{da}{dt} \tag{3.19b}$$

such that the final equation relating the integral of a derivative to the derivative of an integral is obtained as:

Leibnitz' Rule

$$\int_{a(t)}^{b(t)} \left.\frac{\partial f(x,t)}{\partial t}\right|_x dx = \frac{d}{dt}\int_{a(t)}^{b(t)} f(x,t)\,dx - f(b,t)\frac{db}{dt} + f(a,t)\frac{da}{dt} \tag{3.20}$$

Note that although this equation has been derived here in the physical context of a generalized function dependent on one spatial coordinate and time, the resulting Leibnitz' rule is valid when x and t refer to any two independent variables.

3.2 GENERALIZED FUNCTIONS FOR A SPACE CURVE

The concept of a generalized function for a straight line introduced in Section 3.1 can readily and naturally be extended to a generalized function for a space curve. If l denotes the coordinate along the curve, then γ^c can be defined in a manner analogous to γ^L in equation (3.2) as:

$$\gamma^c(l,t) = H[g^c(l;a)] - H[g^c(l;b)] \quad \textbf{(3.21a)}$$

such that:

$$\gamma^c(l,t) = \begin{cases} 0 & l < a(t) \\ 1 & a(t) < l < b(t) \\ 0 & l > b(t) \end{cases} \quad \textbf{(3.21b)}$$

The function $g^c(l;a)$ relates to the position along the curve and is a conceptual extension of the function $x - a$ used in equation (3.2). As such, it will be referred to as the position function. Note that g^c depends on the independent variable as well as a parameter defining a point on the curve. Likewise H is a conceptual extension of the step function U given by equation (3.1a) and undergoes a step change in value where the position function g^c is equal to zero. The superscript c is used to indicate that the generalized function is defined along a simple space curve. Time dependence of γ^c occurs through the dependence of a and b on time. Figure 3.4 illustrates use of γ^c to delineate a portion of a space curve.

3.2.1 Curvilineal Derivatives of γ^c

In Section 3.1, the generalized function γ^L was used to identify a portion of a straight line in one dimensional space. Thus the orientation of the line with respect to a cartesian coordinate system did not change as a function of position. Calculation of the spatial derivative of γ^L simply required differentiation with respect to a cartesian coordinate aligned parallel to the line. For a space curve, however, the orientation is not necessarily constant so the calculation of the spatial derivative requires that the curvature be accounted for. If $\boldsymbol{\lambda}$ is a unit vector tangent to a curve C, the change in γ^c with position along the curve is $\boldsymbol{\lambda} \cdot \nabla^c \gamma^c$. In terms of γ^c defined in equation (3.21a):

$$\boldsymbol{\lambda} \cdot \nabla^c \gamma^c = \delta[g^c(l;a)] - \delta[g^c(l;b)] \quad \textbf{(3.22)}$$

If $\mathbf{e}$ is defined to be a unit vector tangent to $\boldsymbol{\lambda}$ but pointing outward from the region of interest $a < l < b$ such that $\mathbf{e} = -\boldsymbol{\lambda}$ at $l = a$ and $\mathbf{e} = \boldsymbol{\lambda}$ at $l = b$ then:

$$\nabla^c \gamma^c = -\mathbf{e}|_{l=a} \delta[g^c(l;a)] - \mathbf{e}|_{l=b} \delta[g^c(l;b)] \quad \textbf{(3.23)}$$

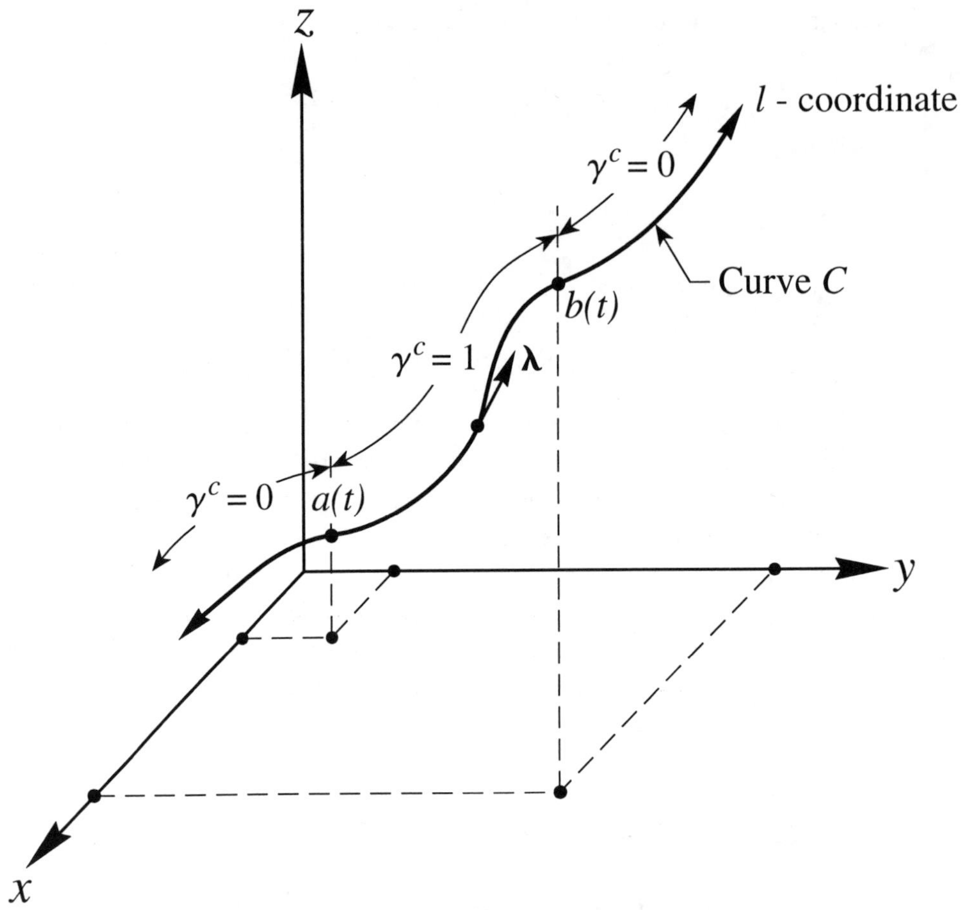

Figure 3.4. Illustration of the use of γ^c to identify a portion of a space curve that is of interest (i.e. segment where $\gamma^c = 1$ is the portion of interest).

In some instances, a generalized function γ^c will be used to indicate a number of segments along a curve. If the N points across which the function undergoes a change in value from 1 to 0 (or vice versa) are indicated as a_i then:

$$\nabla^c \gamma^c = -\sum_{i=1}^{N} \mathbf{e}\delta[g^c(l;a_i)] \tag{3.24}$$

where the dependence of the sense of **e** on a_i is understood. For the case where the curve of interest is a closed curve such that $\gamma^c = 1$ along the entire curve (i. e. $N = 0$ in equation (3.24)), $\nabla^c \gamma^c = 0$ along the entire curve.

3.2.2 Integral Properties of $\nabla^c \gamma^c$

Evaluation of an integral over a curve C when the integrand contains $\nabla^c \gamma^c$ is a direct extension of equation (3.9). Specifically, if $\nabla^c \gamma^c$ is defined by equation (3.23) and a function f is continuous in the domain $a < l < b$ then:

$$-\int_{-\infty}^{\infty} f \boldsymbol{\lambda} \bullet \nabla^c \gamma^c dC = f|_{l=b} - f|_{l=a} \tag{3.25}$$

For the more general case where γ^c is used to identify more than one segment of a curve C such that equation (3.24) applies, the integral becomes:

$$-\int_{-\infty}^{\infty} f \boldsymbol{\lambda} \bullet \nabla^c \gamma^c dC = \sum_{i=1}^{N} (\boldsymbol{\lambda} \bullet \mathbf{e} f)|_{l=a_i} \tag{3.26a}$$

or:

$$-\int_{-\infty}^{\infty} f \nabla^c \gamma^c dC = \sum_{i=1}^{N} (\mathbf{e} f)|_{l=a_i} \tag{3.26b}$$

Recall that $\mathbf{e}$ points outward from the region identified by $\gamma^c = 1$ at each of the points a_i such that $\mathbf{e}$ will equal either $\boldsymbol{\lambda}$ or $-\boldsymbol{\lambda}$.

3.2.3 Time Derivative of γ^c

The generalized function γ^c depends on the coordinate along the curve, l, as well as time. The total time derivative of γ^c along the curve is given by:

$$\frac{d\gamma^c}{dt} = \left.\frac{\partial \gamma^c}{\partial t}\right|_l + \frac{dl}{dt} \boldsymbol{\lambda} \bullet \nabla^c \gamma^c \tag{3.27}$$

Similarly to the straight line case, this total derivative will be zero if dl/dt is set equal to an observation velocity along the curve, $\mathbf{w}^c$, under the constraints:

1. $\mathbf{w}^c$ is tangent to the curve and at the boundary of the portion of the curve of interest is equal to the velocity of that boundary; and
2. at other points on the curve, $\mathbf{w}^c$ is such that the boundary of the portion of the curve of interest is never crossed.

Under these constraints:

$$\left.\frac{\partial \gamma^c}{\partial t}\right|_l = -\mathbf{w}^c \bullet \nabla^c \gamma^c \tag{3.28}$$

Equation (3.28) is useful in that the partial time derivative of γ^c may be expressed in terms of spatial derivatives with integral properties as described in Section 3.2.2. As a final note, it should be pointed out that if the curve of interest is a closed curve, both $\nabla^c \gamma^c$ and $\partial \gamma^c / \partial t|_l$ will be zero.

3.3 GENERALIZED FUNCTIONS FOR A SURFACE

Consider a general, orientable, not necessarily planar surface in space. Any position on the surface can be described in terms of two orthogonal surface coordinates which are denoted as $\mathbf{u} = (u_1, u_2)$. In turn, $\mathbf{u}$ will depend on the spatial coordinates $\mathbf{x} = (x, y, z)$. A generalized function γ^s can now be defined in terms of $\mathbf{u}$ and time to delineate any region on the surface as:

$$\gamma^s(\mathbf{u}, t) = \begin{cases} 1 \text{ on the portion of the surface which is of interest} \\ 0 \text{ on the remainder of the surface} \end{cases} \tag{3.29}$$

In this manuscript, "surfaces of interest" will be restricted to a finite number of non-overlapping simple regions. Figure 3.5 illustrates use of γ^s to define a region on a surface. The dependence of γ^s on time accounts for the expansion or contraction of the region of interest on the entire surface.

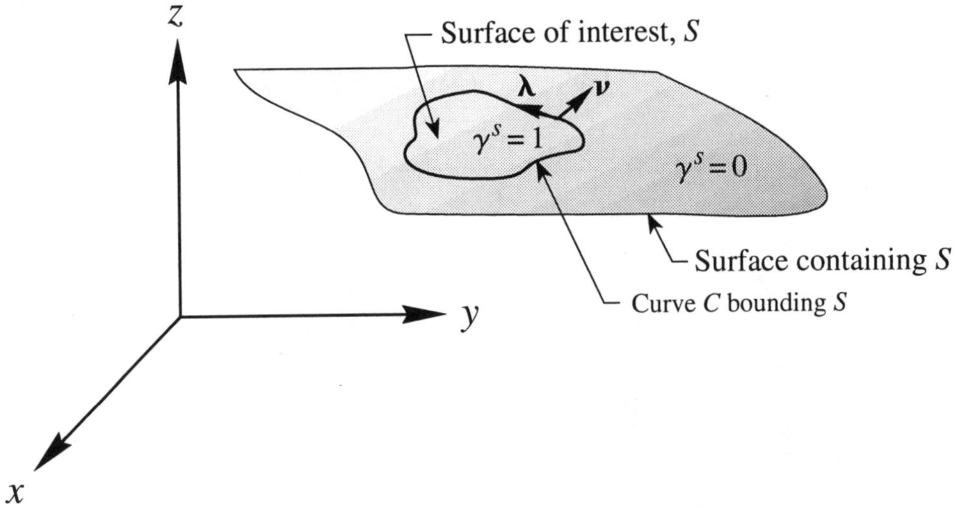

Figure 3.5. Illustration of the use of γ^s to identify a portion of a surface that is of interest (i.e. portion of surface indicated as S where $\gamma^s = 1$ is of interest).

3.3.1 Surficial Derivative of γ^s

From the definition of a surface gradient, $\nabla^s \gamma^s$ is a vector pointing in the direction of greatest change in γ^s, i.e. normal to the region defined by $\gamma^s = 1$. For a surface, where position is specified using two orthogonal surface coordinates, the surficial gradient of γ^s is zero everywhere on the surface except along the curve C dividing the portion of the surface that is of interest from the extension of that surface. Accordingly, the surficial gradient of γ^s is given by:

$$\nabla^s \gamma^s = -\boldsymbol{\nu}\delta[g^s(\mathbf{u};\mathbf{u}^c)] \tag{3.30}$$

where $\boldsymbol{\nu}$ is a unit vector tangent to the surface and outwardly normal to the region where $\gamma^s = 1$ (see figure 3.5). The points $\mathbf{u}^c$ lie on the curve C; and the position function $g^s(\mathbf{u};\mathbf{u}^c) = 0$ defines curve C in terms of the surface coordinate $\mathbf{u}$. If γ^s is used to identify N separate portions of a surface then equation (3.30) can be extended to the form:

$$\nabla^s \gamma^s = -\sum_{i=1}^{N} \boldsymbol{\nu}\delta[g_i^s(\mathbf{u};\mathbf{u}_i^c)] \tag{3.31}$$

where the subscript i is used to denote each of the surfaces of interest.

The similarity between this equation and equation (3.24) for a curve should be apparent. Furthermore, when the surface of interest is a closed surface such that $\gamma^s = 1$ along the entire surface (i. e. $N = 0$ in equation (3.31) because no boundary curve exists), $\nabla^s \gamma^s = 0$ along the entire surface.

3.3.2 Integral Properties of $\nabla^s \gamma^s$

In Section 3.2.2 concerning integration over a curve, it was shown that when $\nabla^c \gamma^c$ appears in the integrand, the integral is converted to a summation of the other terms in the integrand evaluated at the endpoints of the curve segments of interest (see equations (3.25) and (3.26a)). In a conceptually analogous manner, when $\nabla^s \gamma^s$ appears in an integral over a large surface that includes all the surficial regions of interest, the integral over the surface may be converted to an integral over the boundary curves of the regions. If γ^s is used to identify one section of a very large surface, S_∞, and the boundary curve of this section is C then:

$$-\int_{S_\infty} f\boldsymbol{\nu}\cdot\nabla^s \gamma^s dS = \int_C f dC \tag{3.32}$$

The use of an integral on the right side of equation (3.32) may be thought of as a natural extension of the summation that appears in equation (3.26a). In that equation, a finite number of points were considered where $\nabla^c \gamma^c$ was non-zero. In equation (3.32) an infinite number of points are considered (i.e. the locus of the curve C) where $\nabla^s \gamma^s$ is non-zero. Hence the summation is replaced by an integration.

If γ^s is used to identify N separate regions of the surface S_∞ where region i is bounded by curve C_i, equation (3.32) takes the form:

$$-\int_{S_\infty} f\boldsymbol{\nu} \cdot \nabla^s \gamma^s dS = \sum_{i=1}^{N} \int_{C_i} f dC \qquad \textbf{(3.33a)}$$

or:

$$-\int_{S_\infty} f \nabla^s \gamma^s dS = \sum_{i=1}^{N} \int_{C_i} f\boldsymbol{\nu} dC \qquad \textbf{(3.33b)}$$

For convenience in notation, the summation over the boundary curves of interest, C_i, will not be explicitly given in subsequent usage. Rather, integration over C will indicate that all boundary curves where $\nabla^s \gamma^s \neq 0$ are included in the integration.

3.3.3 Time Derivative of γ^s

The generalized function γ^s depends on the surface coordinates, $\mathbf{u}$, as well as time. The total time derivative of γ^s in the surface may be written:

$$\frac{d\gamma^s}{dt} = \left.\frac{\partial \gamma^s}{\partial t}\right|_{\mathbf{u}} + \frac{d\mathbf{u}}{dt} \cdot \nabla^s \gamma^s \qquad \textbf{(3.34)}$$

This total derivative will be zero if $d\mathbf{u}/dt$ is specified to be the observation velocity in the surface, $\mathbf{w}^s$, constrained such that:

1. the observation velocity, $\mathbf{w}^s$, on the boundary between the portion of the surface of interest and the extended surface (i. e. where $g^s(\mathbf{u};\mathbf{u}^c) = 0$) equals the velocity of the boundary; and
2. the velocity of the observer at other locations on the surface is such that the boundary curve of the portion of the surface of interest is never crossed.

Under these constraints, no change in the value of γ^s will be observed so $d\gamma^s/dt = 0$ and:

$$\frac{\partial \gamma^s}{\partial t}\bigg|_{\mathbf{u}} = -\mathbf{w}^s \cdot \nabla^s \gamma^s \qquad (3.35)$$

This equation relates the time derivative evaluated at a point on the surface to spatial derivatives with known integral properties. Both of the above constraints are satisfied by all theorems developed in this work.

3.4 GENERALIZED FUNCTIONS FOR A VOLUME

In three-dimensional space, any point may be located by specifying, for example, its Cartesian coordinates $\mathbf{x} = (x, y, z)$. A spatial generalized function, denoted as γ (without superscript) can then be defined in terms of the global coordinates and time. This function can be employed to identify any region of all space by requiring that:

$$\gamma(\mathbf{x}, t) = \begin{cases} 1 \ \text{ in the spatial region of interest} \\ 0 \ \text{ in the rest of space} \end{cases} \qquad (3.36)$$

The region of interest may consist of one or more non-overlapping finite volumes, each bounded by an orientable surface. The dependence of γ on time accounts for the expansion, contraction, and deformation of the spatial region of interest. Figure 3.6 illustrates use of γ to identify a region in space.

3.4.1 Spatial Derivative of γ

The spatial gradient of γ will be a vector pointing in the direction of greatest change in γ. Within and exterior to the region of interest, γ is constant so that $\nabla \gamma$ will be 0. At the boundary of the region of interest, the direction of greatest change in γ is normal to the boundary and the magnitude of the change is characterized by the Dirac delta function. For a simple volume with boundary S, the gradient of γ is

$$\nabla \gamma = -\mathbf{n}^* \delta[g(\mathbf{x};\mathbf{x}^s)] \qquad (3.37)$$

where $\mathbf{n}^*$ is the outwardly directed unit vector normal to the region where $\gamma = 1$ (see figure 3.6); the coordinates $\mathbf{x}^s$ are on the surface S; and $g(\mathbf{x};\mathbf{x}^s) = 0$ defines surface S in terms of the spatial coordinates. If the spatial generalized function is used to identify N distinct regions in space, then equation (3.37) can be extended to:

$$\nabla \gamma = -\sum_{i=1}^{N} \mathbf{n}^* \delta[g_i(\mathbf{x};\mathbf{x}_i^s)] \qquad (3.38)$$

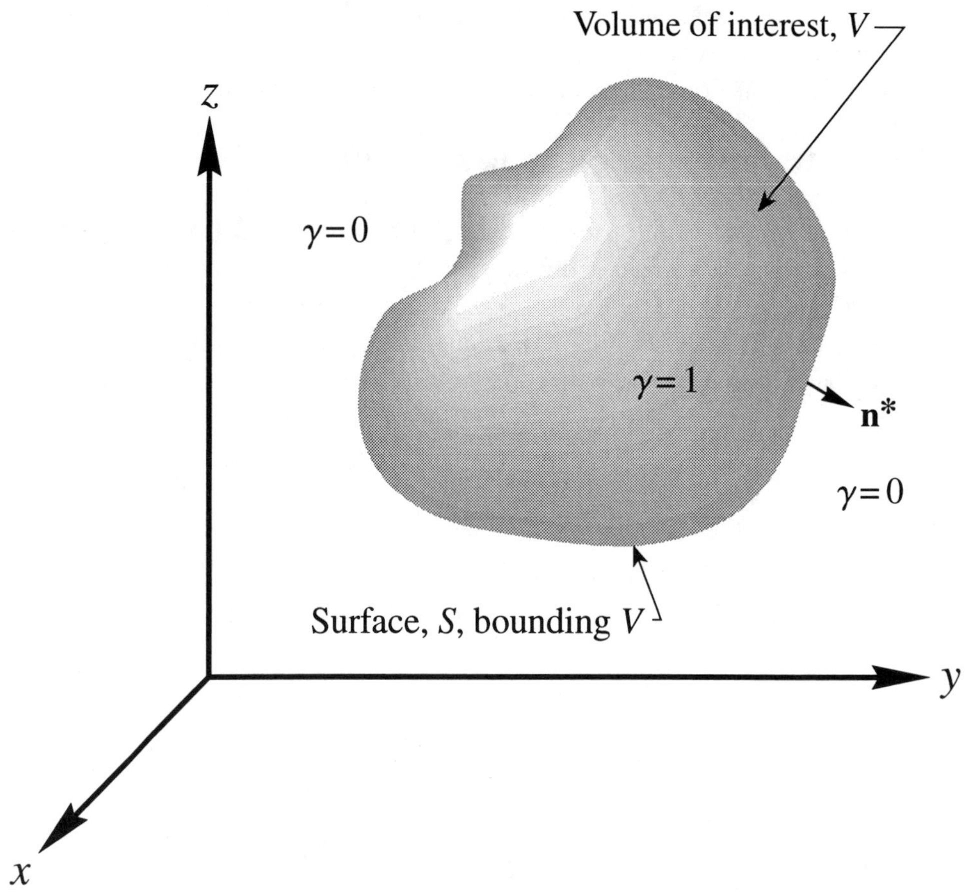

Figure 3.6. Illustration of the use of γ to identify a portion of space that is of interest (i.e. shaded region designated as V where $\gamma = 1$ is the portion of all space of interest).

where i is used to designate each of the regions of interest and the unit normal vector $\mathbf{n}^*$ varies with position on the surface. The similarity of this equation to the expression for $\nabla^c \gamma^c$ in equation (3.24) and to the expression for $\nabla^s \gamma^s$ in equation (3.31) is apparent.

3.4.2 Integral Properties of $\nabla \gamma$

The integral properties of $\nabla \gamma$ may be obtained as a simple extension of the reasoning employed in Section 3.3.2 in considering $\nabla^s \gamma^s$. When $\nabla \gamma$ appears in the integrand of a volume integral over all space, the integral converts to an integral over the surface of the region of interest. If γ is used to identify one volume, V, which is contained in V_∞ and has bounding surface S, then:

$$-\int_{V_\infty} f\mathbf{n}^* \bullet \nabla \gamma dV = \int_S f dS \tag{3.39}$$

If γ is used to identify N separate regions of the volume V_∞, as in equation (3.38), where S_i is the boundary of volume V_i, equation (3.39) may be written as:

$$-\int_{V_\infty} f\mathbf{n}^* \bullet \nabla \gamma dV = \sum_{i=1}^{N} \int_{S_i} f dS \tag{3.40a}$$

or:

$$-\int_{V_\infty} f \nabla \gamma dV = \sum_{i=1}^{N} \int_{S_i} f\mathbf{n}^* dS \tag{3.40b}$$

In subsequent usage, the summation over i and subscript on S will be omitted. Integration over a surface S will be understood to include all boundary surfaces of volumes in V_∞ identified using the generalized function γ.

3.4.3 Time Derivative of γ

The generalized function γ depends on the spatial coordinates $\mathbf{x}$ as well as time. Hence the total time derivative takes the form:

$$\frac{d\gamma}{dt} = \left.\frac{\partial \gamma}{\partial t}\right|_{\mathbf{x}} + \frac{d\mathbf{x}}{dt} \bullet \nabla \gamma \tag{3.41}$$

This total derivative will be zero if the observation velocity, $d\mathbf{x}/dt$, is set equal to $\mathbf{w}$ where:

1. the velocity of the observer, $\mathbf{w}$, on the boundary S equals the velocity of the boundary (i. e. an observer located at $g(\mathbf{x};\mathbf{x}^s) = 0$ remains at a point where this equation is satisfied); and
2. the velocity of an observer not on the boundary S is such that the boundary is never crossed resulting in the observer seeing no change in γ.

With these constraints, $d\gamma/dt$ will be zero and:

$$\left.\frac{\partial \gamma}{\partial t}\right|_{\mathbf{x}} = -\mathbf{w} \bullet \nabla \gamma \tag{3.42}$$

which relates time derivatives to spatial derivatives with known integral properties. In this work, the above constraints on $\mathbf{w}$ are always imposed.

3.5 FURTHER DISCUSSION OF GENERALIZED FUNCTIONS

3.5.1 A Look at Other Works

Derivative and integral properties of generalized functions presented in Sections 3.2 - 3.4 are developed from a somewhat heuristic viewpoint to allow a clearer understanding of important relations while minimizing mathematical complexities. Other means of arriving at the same conclusions exist in the literature.

Farassat [1977] presents generalized functions by defining them in terms of a continuous linear functional over a given function space. He then defines derivatives of generalized functions in terms of the functional in both one-dimension and higher dimensions and explores some of their integral properties. All results presented herein are consistent with his results.

Some insight into the gradient of γ in three dimensions (defined in equation (3.37) for a single volume) can be gained by looking at this expression in the context of Farassat's formalism. In *Farassat* [1977] and in *Kanwal* [1983], the following relations appear (with some notation modified to conform with the present formalism):

$$\overline{\nabla} F = \nabla F + (\Delta F)(\nabla g)(\delta[g(\mathbf{x};\mathbf{x}^s)]) \tag{3.43}$$

where

$\overline{\nabla} F$ indicates the generalized or distributional gradient of a generalized function F,

∇F is the derivative of the smooth part of F,

$g(\mathbf{x};\mathbf{x}^s) = 0$ defines a surface across which F experiences a jump in value,

$\delta[g(\mathbf{x};\mathbf{x}^s)]$ is the Dirac delta function that is non-zero on the surface where $g(\mathbf{x};\mathbf{x}^s) = 0$, and

ΔF is the jump in F at $g(\mathbf{x};\mathbf{x}^s) = 0$.

To relate the expression given in equation (3.43) to the derivation in Section 3.4, let F be the generalized function formed as a product of some functions f_{I} and f_{II} with the generalized functions γ_{I} and γ_{II} used to distinguish between two regions of space separated by the surface $g(\mathbf{x};\mathbf{x}^s) = 0$. Associate γ_{I} with region I such that $\gamma_{\mathrm{I}} = 1$ in region I but is zero in region II. Similarly associate γ_{II} with region II such that $\gamma_{\mathrm{II}} = 1 - \gamma_{\mathrm{I}}$ in both regions of space. Define f_{I} to equal F in region I but extend smoothly into region II while f_{II} is equal to F in region II but extends smoothly into region I. Since f_{I} and f_{II} are both smooth, the discontinuity in F is accounted for by application of γ_{I} and γ_{II} such that:

$$F = f_{\text{I}}\gamma_{\text{I}} + f_{\text{II}}\gamma_{\text{II}} \tag{3.44}$$

Then use of the standard gradient notation and application of the chain rule yields:

$$\nabla F = \gamma_{\text{I}}(\nabla f_{\text{I}}) + \gamma_{\text{II}}(\nabla f_{\text{II}}) + f_{\text{I}}(\nabla\gamma_{\text{I}}) + f_{\text{II}}(\nabla\gamma_{\text{II}}) \tag{3.45}$$

Now make use of equation (3.37) to eliminate the gradients of the generalized functions, and note that $\mathbf{n}_{\text{I}}{}^* = -\mathbf{n}_{\text{II}}{}^*$ to obtain:

$$\nabla F = \gamma_{\text{I}}(\nabla f_{\text{I}}) + \gamma_{\text{II}}(\nabla f_{\text{II}}) + \mathbf{n}_{\text{I}}{}^*(f_{\text{II}} - f_{\text{I}})\,\delta[g(\mathbf{x};\mathbf{x}^s)] \tag{3.46}$$

The term on the left corresponds to $\overline{\nabla}F$ in equation (3.43), the first two terms on the right account for the gradient of the smooth part of F, and the last term accounts for the discontinuity in F at the boundary between the two domains. Thus equations (3.43) and (3.46) are equivalent representations of the same concept.

Kinnmark and Gray [1984] provide an alternative derivation of the results presented in Section 3.4 in their exposition on generalized functions. Their proof of equation (3.37) starts by defining $\gamma(\mathbf{x}, t)$ in terms of a limit of a sequence of functions:

$$\gamma(\mathbf{x}, t) = \frac{1}{2} + \frac{1}{2}\lim_{m\to\infty}\{\text{erf}[-\sqrt{m}(\mathbf{x} - \mathbf{x}^s)\bullet\mathbf{n}]\} \tag{3.47}$$

Note that this definition is a multidimensional analogue of equations (3.1). By differentiating equation (3.47) and using concepts from the theory of distributions, *Kinnmark and Gray* [1984] obtain equation (3.37). In the same paper, *Kinnmark and Gray* [1984] provide a rigorous development of the integral property of γ for convex simple regions that results in an expression identical to equation (3.39).

3.6 CONCLUSION

The objective of this chapter has been to provide an exposition of generalized functions used to identify volumes, surfaces, and curves. The integral properties and relations between the temporal and spatial derivatives of these functions may be exploited to allow for relatively simple derivations of integration and averaging theorems. Another feature of these functions which is very important is their utility in interrelating integrals over curves, surfaces, and volumes. In the Chapters 4 and 6, some useful identities will be presented which complete the understanding needed to derive the theorems which form the primary subject of this text.

3.7 REFERENCES

Dirac, P. A. M., *The Principles of Quantum Mechanics*, Clarendon, Oxford, 1930.

Farassat, F., Discontinuities in aerodynamics and aeroacoustics: the concept and applications of generalized functions, *J. Sound and Vibration*, **55**(2), 165-193, 1977.

Gel'fand, I. M., and G. E. Shilov, *Generalized Functions Properties and Operations*, Vol. 1, Academic, New York, 1964.

Gray, W. G., and M. A. Celia, On the use of generalized functions in engineering analysis, *Intl. J. of Applied Engineering Education*, **6**(1), 89-96, 1990.

Hildebrand, F. B., *Advanced Calculus for Applications*, Second edition, Prentice-Hall, Englewood Cliffs, 1976.

Kanwal, R. P., *Generalized Functions - Theory and Technique*, Academic, New York, 1983.

Kinnmark, I. P. E., and W. G. Gray, An exposition of the distribution function used in proving the averaging theorems for multiphase flow, *Advances in Water Resources*, **7**(3), 113-115, 1984.

Lighthill, M. J., *Fourier Analysis and Generalised Functions*, Cambridge University, Cambridge, 1958.

Schwartz, L., Théorie des distributions, *Actualités Scientifique et Industrielles*, **1091**(I), Hermann, Paris, 1950.

Schwartz, L., Théorie des distributions, *Actualités Scientifique et Industrielles*, **1122**(II), Hermann, Paris, 1951.

Slattery, J. C., *Momentum, Energy, and Mass Transfer in Continua*, McGraw-Hill, New York, 1972.

Whitaker, S., *Introduction to Fluid Mechanics*, Prentice-Hall, Englewood Cliffs, 1968.

CHAPTER FOUR

INTEGRATION SCALES AND COORDINATE SYSTEMS

4.0 INTRODUCTION

The fundamental purpose of the integration approach employed here is to obtain theorems that describe the transformation of derivatives from one spatial scale to another. This transformation of scale is accomplished by integrating smaller scale equations to obtain forms relevant at larger scales. Integrations of these small scale equations assume a variety of forms depending upon the application. For example, a single averaged value may be desired for an entire region. Alternatively, averaged values for smaller sub-regions within a large region may be needed. The primary difference between these two very different types of averages lies in selection of the length scale for integration. Three spatial scales of integration are considered here. In order of increasing size they are referred to as microscopic, macroscopic, and megascopic scales.

The microscopic scale is defined to be above the level of molecular behavior and is commonly considered as a continuum scale of variation. For example, in the layered aquifer system of figure 4.1, the microscopic scale is used to describe behavior within a single pore. Modeling of flow in an aquifer from a microscopic perspective would require detailed knowledge of the pore geometries and structure throughout the region of interest.

A second scale, the macroscopic scale, may be used to model variation of some average property of interest. Macroscale functions may vary throughout space and with time. In general, transformation to a macroscale perspective may be accomplished by integrating microscale equations and/or properties over some region of space that has a characteristic length much greater than the microscale length but much smaller than the system length scale. For example, in a porous medium, the macroscale is commonly used for

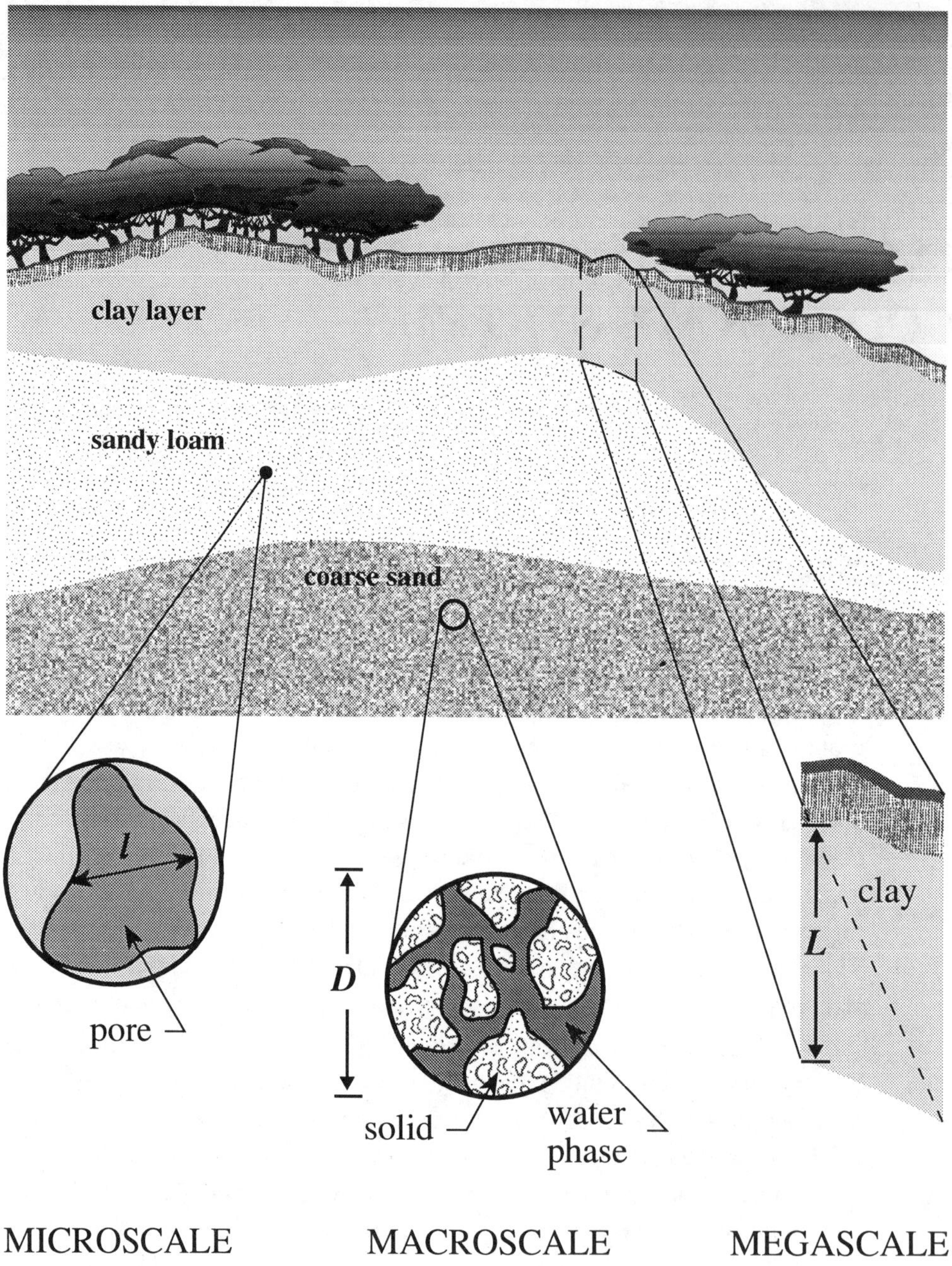

Figure 4.1. A layered aquifer system depicting three spatial scales at which the system may be studied: microscopic or pore scale, macroscopic or representative elementary volume scale, megascopic or system dimension scale.

describing the differential variation of aquifer properties. With reference to figure 4.1, macroscopic scale properties are characterized by measurements taken over a large enough sample of pores that properties measured describe the collective behavior of the pores. For many functions, such as density or velocity, macroscale values are obtained as averages of microscale values. Other macroscopic functions, such as porosity of a porous medium, have no microscopic analogs but rather are quantities that arise at the macroscale in characterizing the system at that scale.

The third scale considered is the megascopic scale. At the megascale, spatial variation may be specified from region to region. The megascopic scale is on the order of the length scale of the system. For the layered aquifer depicted in figure 4.1, the thickness of each geologic layer can be considered as a megascopic scale. To obtain a megascopic variation of some property from layer to layer, one would specify some average property over the thickness of the layer. Note that in this context, megascopic values can be defined by vertical integration while still allowing for microscale or macroscale change in the horizontal directions. A completely megascopic description of a property of a geologic layer would be obtained by specifying one value which characterizes the entire layer. Furthermore, a megascale may be defined that is large enough to encompasses the entire aquifer. Thus, a single value of a function would characterize the entire system. Note that the megascopic length scale will have quite different magnitudes in different directions in systems where, for example, lateral extent is much greater than vertical thickness. A megascopic description of a system is seen to be much simpler than a description at a smaller scale, but this simplification is obtained at the cost of loss of detail in describing physical phenomena which may occur.

As an alternative example to the multi-layer aquifer system, consider flow in a river as shown in figure 4.2. For this case, the turbulent behavior of individual fluid parcels is described at the microscale (Note that in obtaining this description, integration over windows in time is commonly implemented to obtain turbulent stress. This integration may be considered a transformation from a microscopic time scale to a macroscopic time scale.). If the detailed motion of fluid parcels is not of interest, the average motion of a group of fluid parcels may be represented by integration of the equations of fluid mechanics to the macroscopic level. If only variation along the river axis needs to be described, megascopic integration over the river cross-section will provide the appropriate equations. Finally, in some instances, it may only be necessary to describe a single quantity for the entire river (e.g. an effective outflow rate) in order to gain insight into a problem of interest. For this situation, integration to the megascale proceeds over the entire river.

It should be noted that precise identification of each of the length scales for a particular system is very difficult. In fact, no precise criteria exist

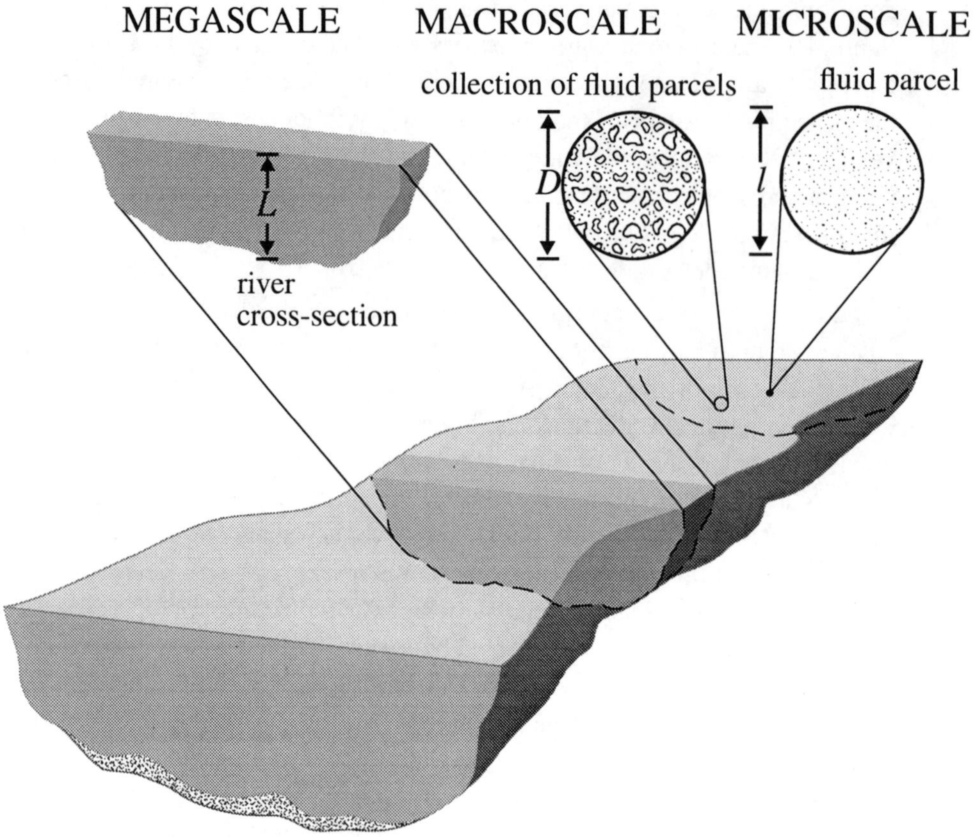

Figure 4.2. River depicting three spatial scales of integration: microscopic or fluid parcel scale, macroscopic scale that is a collection of fluid parcels, and megascopic scale or scale of the river cross-section. Note that as depicted here, the river cross-section is macroscopic in the direction parallel to the channel.

for specification of the various scales. Nevertheless, for a change of scale to lead to a significantly different and mathematically useful description of a phenomenon of interest, the change must involve significantly different length scales. Mathematically, this can be expressed as:

$$l \ll D \ll L \tag{4.1}$$

where l, D, and L are the microscopic, macroscopic, and megascopic length scales, respectively. When equation (4.1) applies, functional variations that occur at a smaller scale may be accounted for by parameterization at a larger scale. Detailed discussions of the three scales of integration are presented subsequently.

4.1 DEFINITION OF INTEGRATION SCALES

The microscopic scale is typically interpreted as the length scale at which the molecular structure of matter is disregarded such that it may be conceptualized as without gaps or empty spaces. Functions describing the system may be defined at this scale that are continuous and have continuous derivatives *Malvern*, 1969]. At this scale, the behavior of individual molecules is accounted for only in an average, continuum sense. In the example of figure 4.1, the microscopic characteristic length, l, is associated with a mean pore diameter. For river flow (figure 4.2), l is related to identification of a collection of fluid molecules such that they form a continuum or parcel.

Equations describing the physics of mass, momentum, and energy transport at the microscopic scale are given by the classic balance laws of continuum mechanics [e. g. as found in *Bird et al.*, 1960; *Eringen*, 1967; *Malvern*, 1969]. Although many microscopic quantities can be measured, modeling the behavior of systems such as an aquifer or river at such a detailed scale is, in some cases, unnecessary or virtually impossible. Thus, the need for measurement and simulation at larger more practical scales arises.

At the largest scale, the megascopic scale, no spatial variations of a property are modeled within the system. Thus, no information is provided about gradients of a property within the interior of a region in the direction of megascopic integration. Instead, the megascale describes what changes occur, on average, in the interior due to conditions imposed at the boundaries of a region (e.g. a change in average concentration due to imposed fluxes of a contaminant at the boundary). The characteristic length of the megascopic scale, L, is defined by the length scale of gross inhomogeneities in the function of interest. In figure 4.1, geologic inhomogeneities exist between layers of the aquifer. Thus, in the vertical direction, the megascopic scale is on the order of the thickness of the aquifer.

The intermediate scale, the macroscopic scale, lies somewhere between the microscopic scale where detailed functional variation is accounted for and the megascopic scale where spatial variations are not accounted for at all. At the macroscale, different phases in a porous medium (e.g. air, water, and solid) are described as overlapping continua, each occupying a fraction of space at a point. However, the characteristic length of the macroscale is not readily defined with its value dependent upon the particular application.

Generally, the macroscopic counterpart of a quantity defined at the microscale is the average of the microscopic quantity over a representative elementary volume (REV). This REV, or integration volume, represents and characterizes a physical point at the macroscale. The characteristic length, D, of the REV is selected such that a macroscopic or average quantity assigned to the REV is insensitive to small changes in length. In an aquifer, such as that depicted in figure 4.1, the characteristic length, D, may be interpreted as the diameter of a volume containing many pores. If D is of the same order of magnitude as the microscopic characteristic length, l, the average quantity may show pronounced fluctuations in its magnitude. Region I in figure 4.3 illustrates the fluctuations of porosity in a porous medium if one tries to characterize it at a microscopic characteristic length scale, l. Likewise, if D approaches the length scale of the megascopic field, L, gross inhomogeneities may produce variations in the value of the averaged quantity. The increase in porosity within region III of figure 4.3 is due to inhomogeneities in the structure of the porous medium. To obtain averaged quantities at the macroscale that are insensitive to small changes in the size of the macroscopic integration volume, a macroscopic characteristic length, D, typically must satisfy equation (4.1). Porosity is a particularly interesting property because it does not actually exist, in any meaningful sense, at the microscale since at this scale a point is either in the voids or in the solid.

In figure 4.3 the porosity remains constant within region II where the characteristic length satisfies the inequality in relation (4.1). In the horizontal direction within each layer, a common approach is to analyze the aquifer at the macroscopic scale. For the example of the layered aquifer in figure 4.1, the macroscopic characteristic length is much larger than that of individual pores yet considerably smaller than the scale of regional changes in geology or flow pattern.

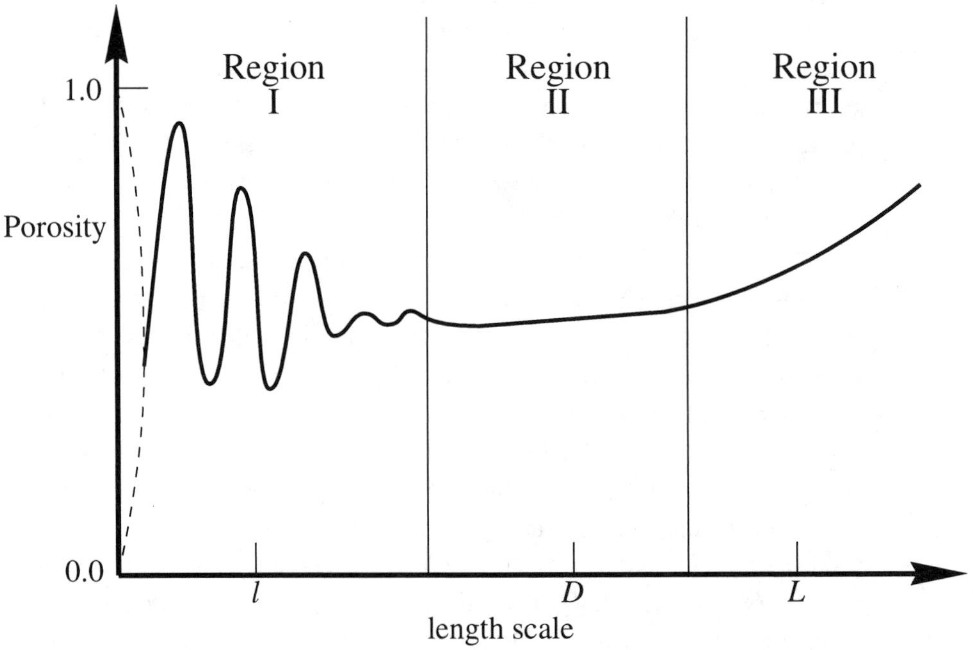

Figure 4.3. Porous medium porosity as a function of measurement length scale.

4.2 COORDINATE SYSTEMS AND INTEGRATION VOLUMES

4.2.1 Integration Theorems

Those theorems referred to here as integration theorems convert integrals of gradient, divergence, curl, and partial time derivatives of a microscopic function to some combination of derivatives of integrals of the microscopic function and integrals over domains of reduced dimensionality. Thus these theorems convert expressions involving the microscale and variations at that scale to expressions involving the microscale and the megascale. Microscopic variations of a function within the domain of megascopic integration are no longer accessible following transformation to the megascale. The megascopic length scale corresponds to the physical dimensions of a problem. To identify a point within the megascopic domain of a problem relevant for the application of an integration theorem, a reference coordinate system is established. This reference, or global, coordinate system is selected to be orthogonal and will be denoted by $\mathbf{x}$ (e.g. $\mathbf{x} = (x, y, z)$ would denote cartesian component directions). The location of the origin of the global system is arbitrary.

Depending on the formulation of the integrand, a megascopic integration region may be selected in equivalent alternative forms. In any direction designated to be megascopic following integration, the integration can be viewed as: 1) performed over an infinite length with a generalized function (presented in detail in Chapter 3) used in the integrand to identify the portion of the domain occupied by the system of interest; or 2) performed only over the domain of the system of interest. The megascopic integration region bounded by the domain of the problem may deform, expand, and contract with time. However, a fully megascopic infinite integration region (i.e. a megascopic region covering all of space) is independent of both space and time in all megascopic directions. For example, by the first approach, when a three-dimensional microscopic function is integrated to the megascale in three coordinate directions using a generalized function to indicate the domain of the system, the integration occurs over all space. This is schematized in figure 4.4 where the generalized function is used to identify the actual system of interest. The origin of the $\mathbf{x}$ coordinate system is arbitrarily located in space. By the second approach, the integration region for conversion to the megascale would not extend infinitely, but would extend only to the boundaries of the domain of interest.

For two-dimensional megascopic integration, the integration region may be either an infinite surface with a generalized function used to identify the portion of the surface constituting the domain of interest or a finite surface whose boundaries coincide with the boundaries of the domain of interest. Because the surface need not be planar, surface coordinates, designated as $\mathbf{u}$

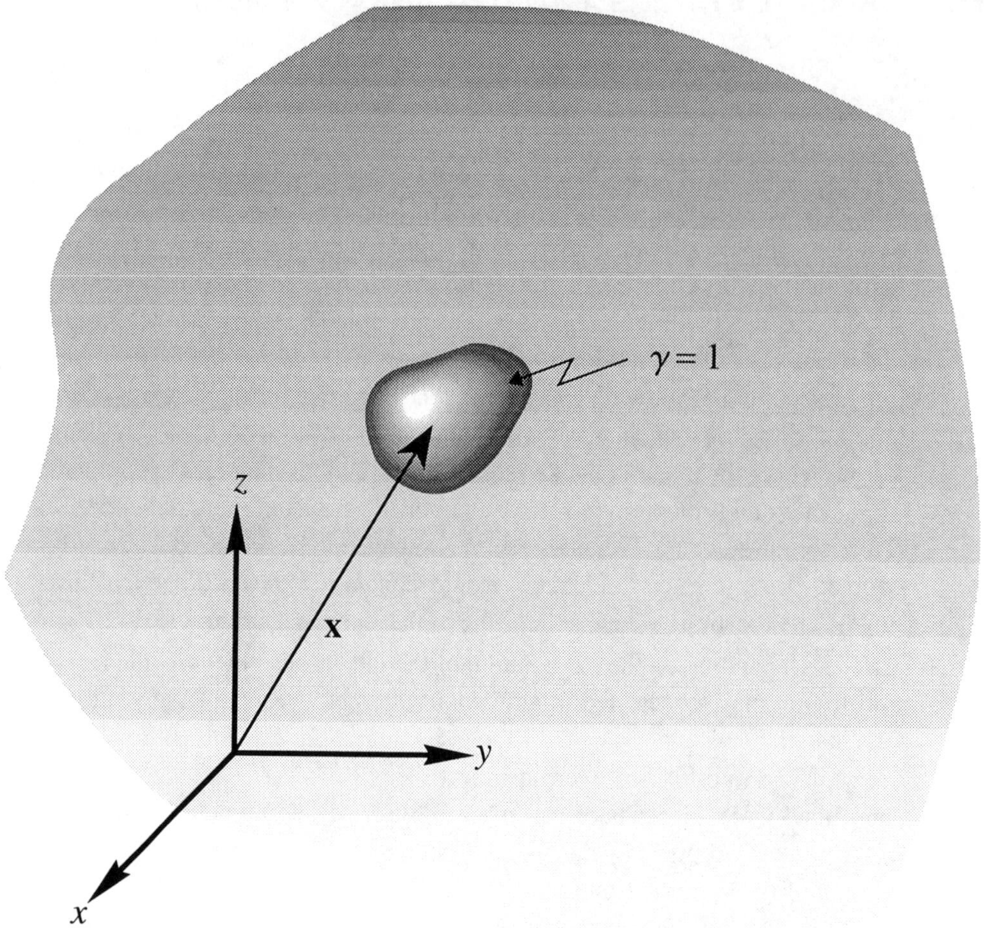

Figure 4.4. A finite volume of interest, the region where $\gamma = 1$, occupying a portion of infinite space.

and defined in terms of the $\mathbf{x}$ coordinate system, are employed. In figure 4.5, the surface is defined designating the two megascopic coordinate directions as λ and ν, and extends infinitely in those directions. A generalized function may be employed to identify the portion of this surface intersecting the domain of interest. In the third coordinate direction, n, the direction normal to the surface, the integrated function retains its microscopic variation.

For one-dimensional megascopic integration, the integration region is simply a curve that may extend to infinity (or perhaps form a large closed loop) when the generalized function is used to identify the portion of the curve within the domain of study. Alternatively, the curve may be finite with its ends coinciding with the physical boundary of the problem of interest. Figure 4.6 illustrates a one-dimensional megascopic integration region along an infinite curve.

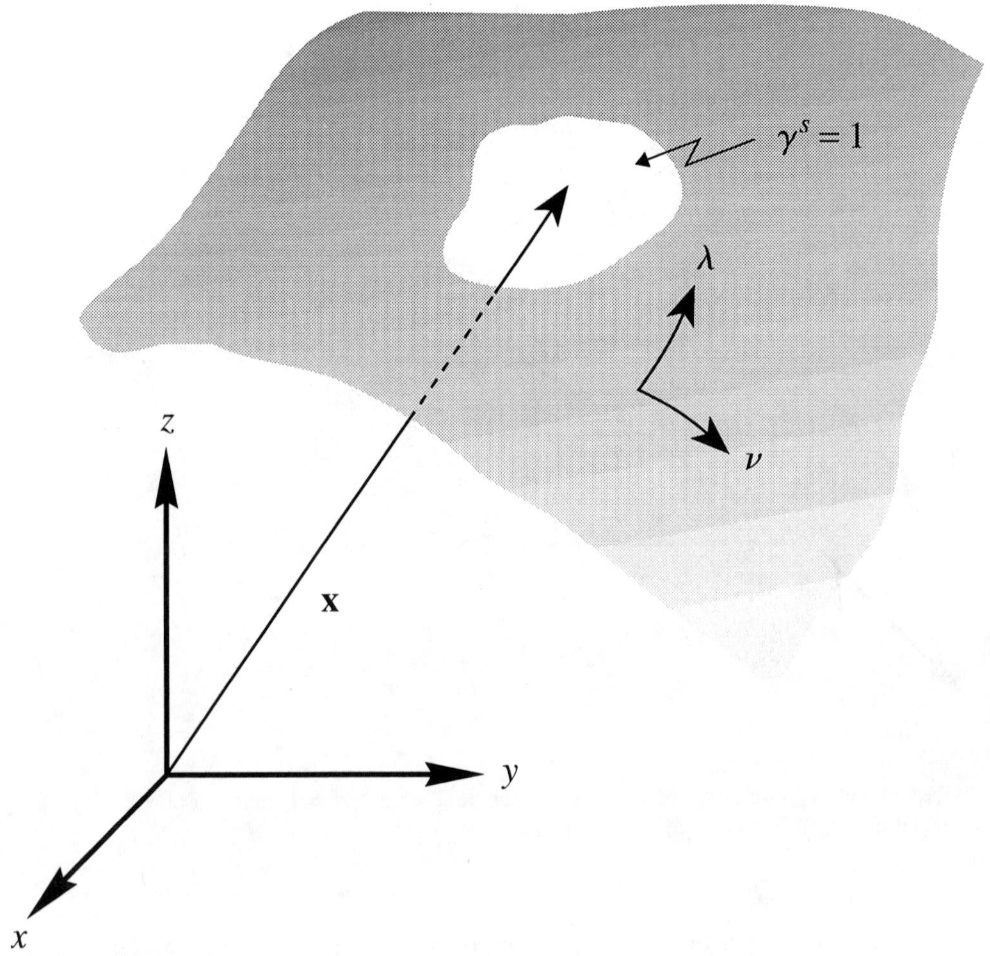

Figure 4.5. A finite surface of interest, the region where $\gamma^s = 1$, occupying a portion of an infinite surface.

The curvilineal coordinate system along the curve is designated as **l** with the single coordinate tangent to the curve indicated as λ. The unit vector $\boldsymbol{\lambda}$ is tangent to λ. The **l** coordinate may be expressed in terms of the global coordinates $\mathbf{x}$. A function integrated to the megascale over the length of the curve would retain its microscopic character in the directions orthogonal to the curve.

4.2.2 Averaging Theorems

In general terms, averaging theorems relate the integral of a derivative to the derivative of that integral. The averaging process transforms microscale equations to equations relevant at some mix of macroscopic and megascopic scales.

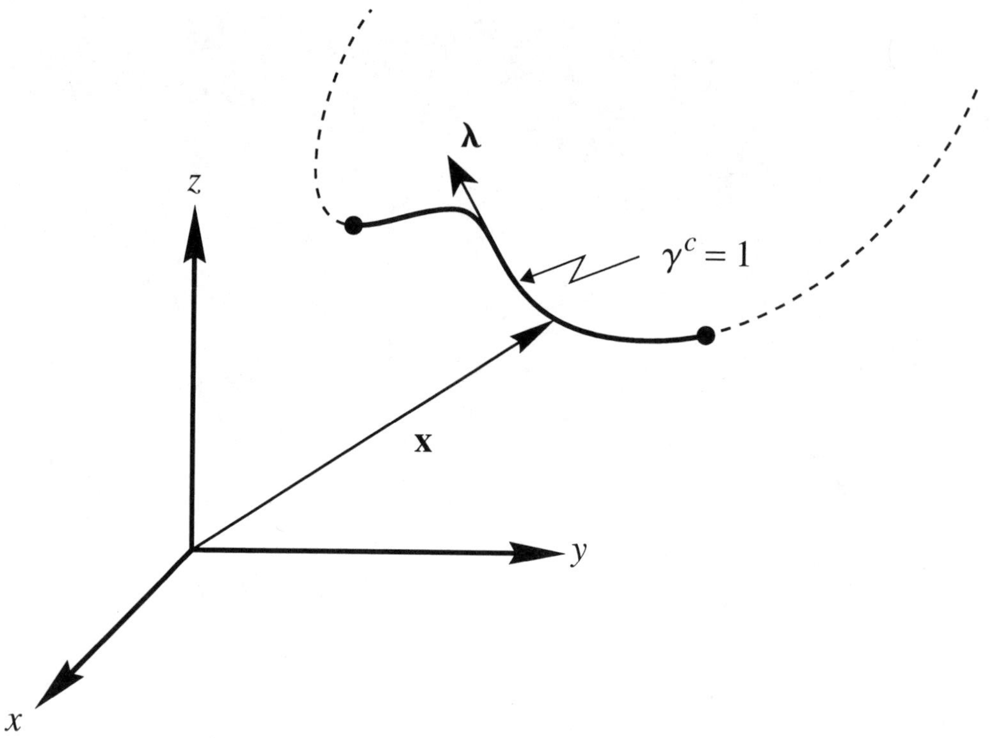

Figure 4.6. A finite curve of interest, the segment where $\gamma^c = 1$, that is a portion of either an infinite or a closed simple curve.

In the derivation of averaging theorems, the integration is performed over a region contained within a volume. These averaging volumes (REV's) are constructed to reflect the different scales of integration for each coordinate direction. Although the averaging region is contained within a volume, the actual integration may occur over curves, surfaces, or subvolumes contained within that REV. After the integration is performed, some average value for the region may be associated with the location of the REV. The global, or $\mathbf{x}$, coordinate system is established in the context of averaging to locate the averaging volume. In this manner, the macroscopic value of the averaged function - even a function defined, for example, using microscopic surficial coordinates within an REV - varies spatially with REV position.

The global coordinate system is used to locate points at the macroscale after averaging. In this manner, a continuous average field may be obtained. Figure 4.7 illustrates different positions, indicated by $\mathbf{x}$, corresponding to different spherical REVs. To account for variation within an REV, a local, or microscale, $\boldsymbol{\xi}$ coordinate system is defined (e.g. with cartesian coordinates designated as ξ, η, and ζ). This system may be defined for each averaging

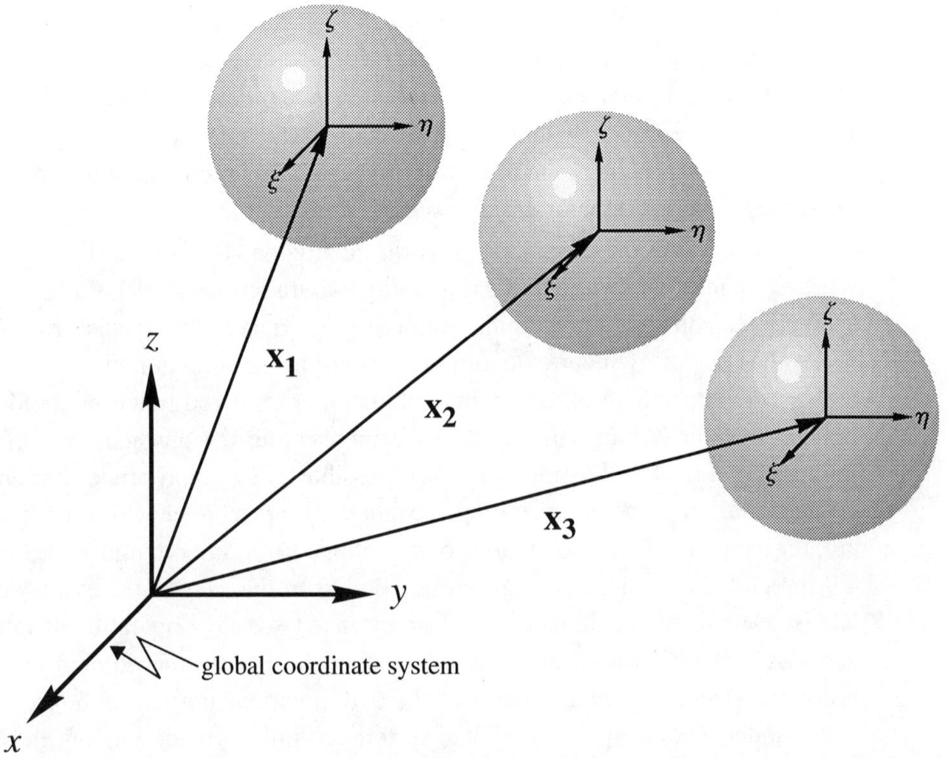

Figure 4.7. Global coordinate system, **x**, used to locate REVs in space and the local coordinate systems, **ξ**, within each REV.

volume. The origin of the local coordinate system is placed at the **x** coordinate that identifies the location of the averaging volume. The local coordinate system associated with each particular REV is illustrated in figure 4.7. Note that the averaging process involves integration over the **ξ** coordinate system, eliminating all microscopic functional dependence on this coordinate, and producing an integrated value of a function associated with the REV position.

For a microscopic function that is completely transformed to the macroscale in all dimensions, the spherical averaging volume depicted in figure 4.7 is appropriate, although any convenient geometry may be selected. The size of these spheres is specified to be independent of time and position. In figure 4.7, the global, **x**, coordinate system identifies the centroid of an REV and the local, **ξ**, coordinate system with origin at this centroid is used in describing variations within the sphere.

Some averaging volumes facilitate a change from the microscale to a combination of macroscale and megascale. In the megascopic dimensions, the averaging volume spans the entire region of interest while in the macroscopic dimensions, the characteristic length, D, satisfies equation (4.1). An averaging region may be implemented either by using it explicitly as the integration region of an integral or by integrating over all space and employing a generalized function in the integrand to identify the averaging region as the portion of all space where the integrand is allowed to be non-zero.

For one-dimensional megascopic averaging, one dimension of the REV is megascopic, while in the remaining two directions the REV is macroscopic. One convenient averaging volume appropriate for this transformation is a right circular cylinder extending to infinity in the direction tangent to its axis as shown in figure 4.8. For this volume, a generalized function could be defined to indicate that portion of space intersecting the physical system of interest. Figure 4.9 illustrates another possible REV appropriate for one-dimensional megascopic averaging: a right circular cylinder whose ends are cut, for example, by the boundaries of a geologic layer. In both figures 4.8 and 4.9 the global coordinate system locates points on the axes of the cylindrical REVs. The axes of the cylinders can be chosen to have any orientation in space and this orientation need not remain constant throughout the domain of the problem. Macroscopic variation is in the two directions normal to the axis of the cylinder. The local, $\boldsymbol{\xi}$, coordinate system has only two coordinates, designated here as ξ and η, which lie in a plane perpendicular to the cylinder axis as illustrated in figures 4.8 and 4.9. This local coordinate system is used to identify microscopic variation around the axis of the cylinder within the REV. The local coordinate system is two-dimensional, requiring no axis parallel to the axis of the cylinder since averaging in that direction is to the megascale.

When megascopic averaging is two-dimensional, only one dimension is transformed to the macroscale. An averaging volume in the shape of an infinite slab, as pictured in figure 4.10, is an appropriate REV for two-dimensional megascopic averaging. With this geometry, the generalized function is used to indicate the portion of the slab that intersects the physical space of interest. Effectively, the REV is equivalent to the slab of figure 4.11 which is bounded, as an example, by the cross-section of a river. In figures 4.10 and 4.11 the faces of a particular averaging slab are parallel. However, the slabs may have different orientations at different positions in space. Macroscopic variation in one dimension is along the axis of the channel as shown in figure 4.10 and figure 4.11. Only one local coordinate, indicated as ξ, is needed to identify variations within the thickness of the slab because that is the only direction where macroscopic variation is being considered. Figure 4.10 illustrates the relation between the local and global coordinate systems for two-dimensional megascopic averaging.

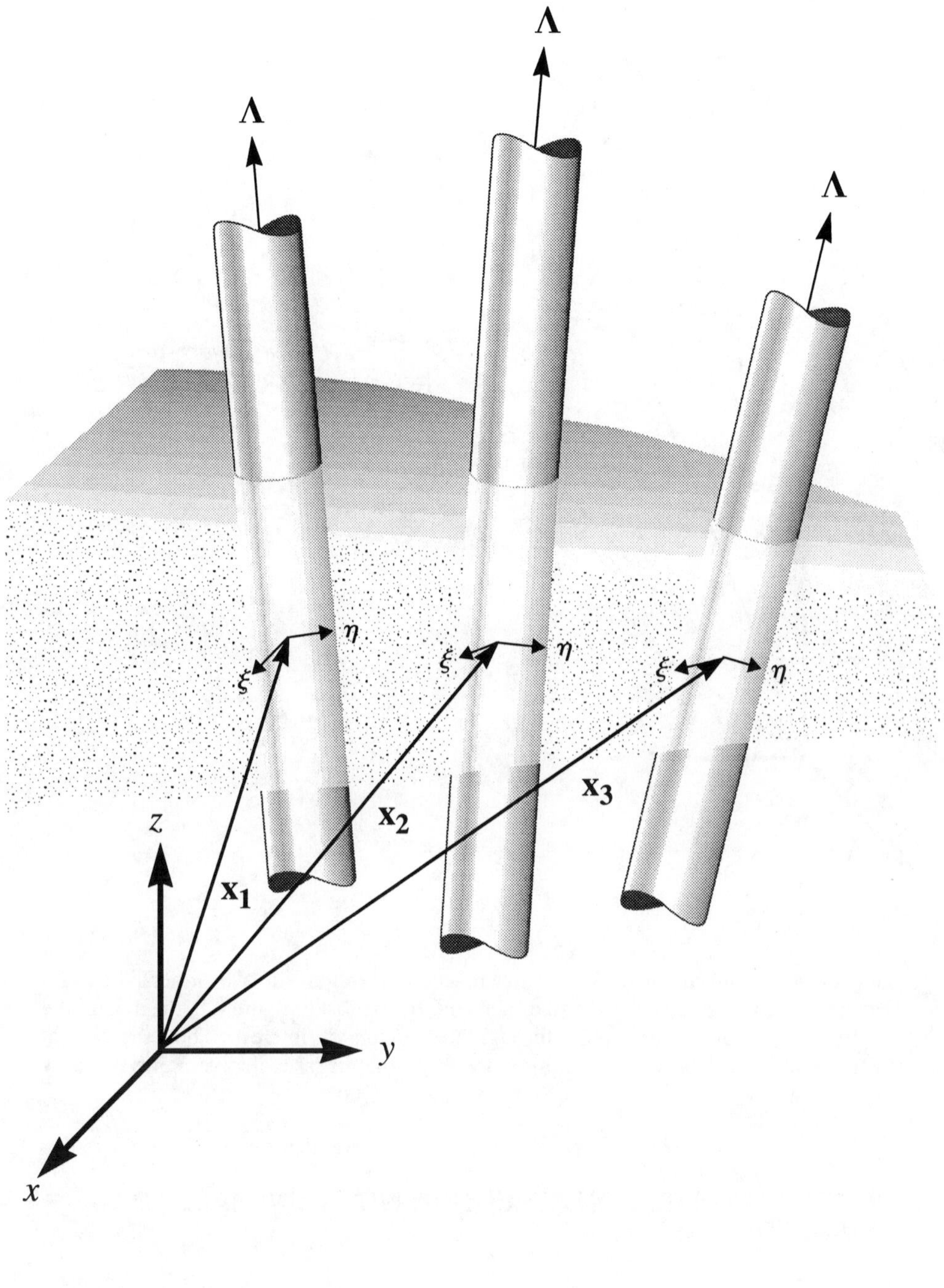

Figure 4.8. Convenient REVs for macroscopic averaging in two dimensions with megascopic averaging in the third dimension. The spatial generalized function γ may be used to distinguish between the part of an REV intersecting the region of interest ($\gamma = 1$) and the parts of the REV that extend to infinity in the megascopic Λ direction ($\gamma = 0$).

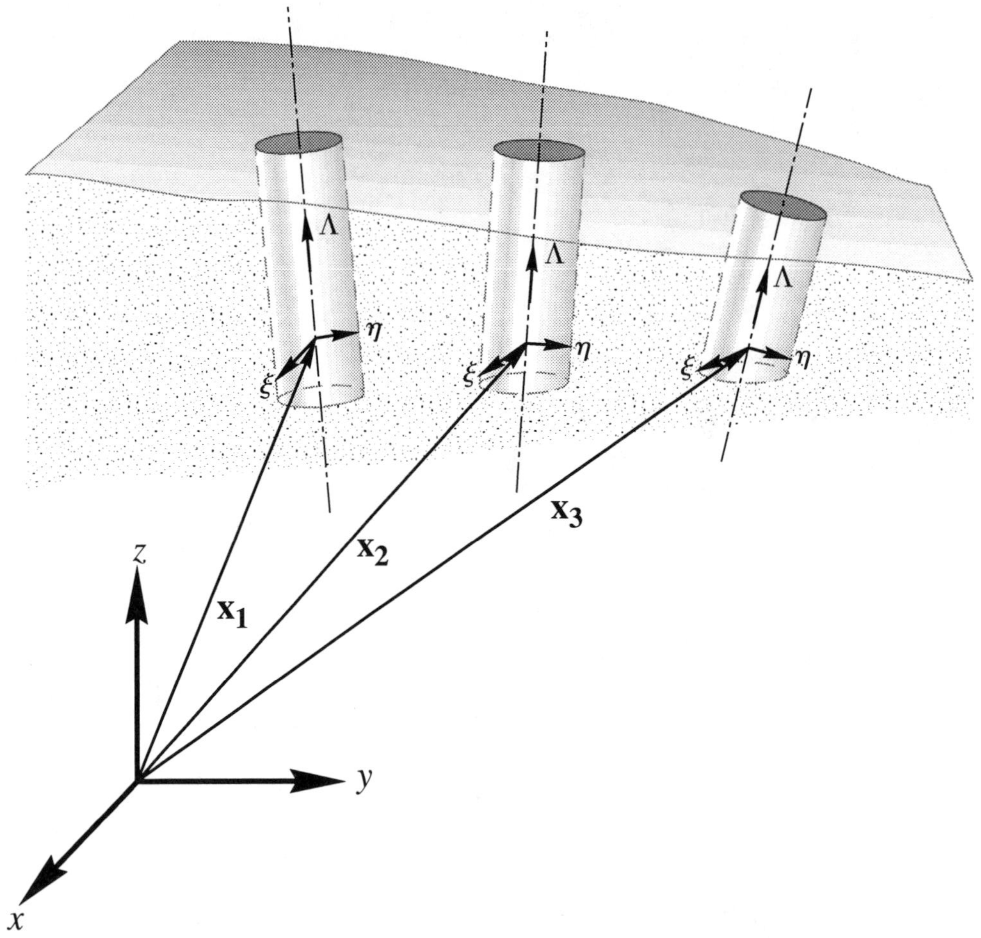

Figure 4.9. Convenient REVs for macroscopic averaging in two dimensions with megascopic averaging in the third dimension. In contrast to figure 4.8, the axis of the cylindrical REVs extend only through the region of interest. Thus the spatial generalized function γ is not required to define the megascopic averaging in the Λ direction.

4.3 INTEGRATION REGION SELECTION WITH GENERALIZED FUNCTIONS

Recall that one motivation behind the use of the generalized functions is to develop a simple method for indicating a volume, surface, or curve of interest. As a consequence of the use of generalized functions, limits of integration that depend on time or space may be replaced by fixed, albeit infinite, limits. This replacement allows the order of differentiation and integration to be reversed, a necessary step in developing averaging and integration theorems. In essence, forming the product of a generalized function with an integrand transfers the

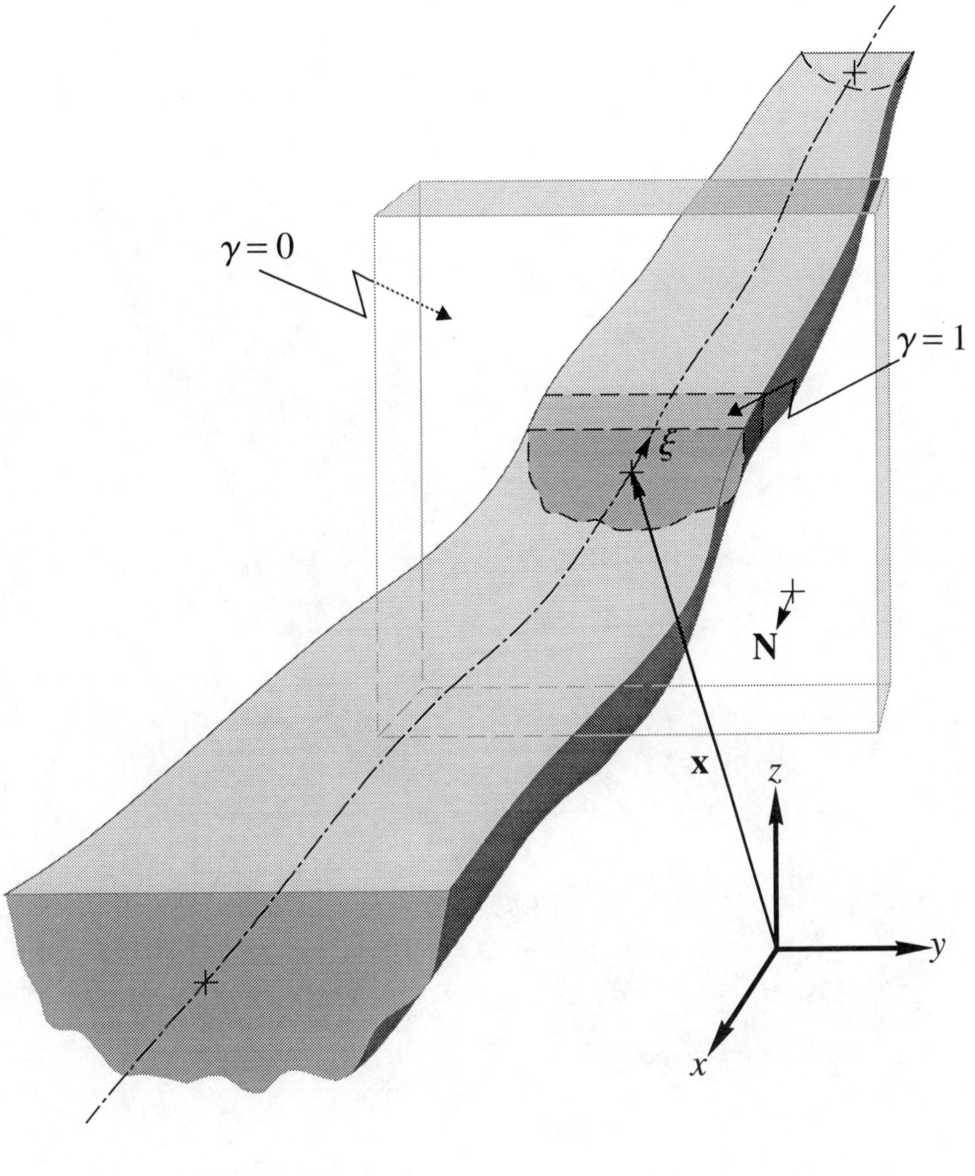

Figure 4.10. Convenient slab REV for macroscopic averaging in one dimension with megascopic averaging in the other two dimensions. The spatial generalized function γ may be used to distinguish between the part of the REV intersecting the region of interest ($\gamma = 1$) and the parts of the REV that extend to infinity in the megascopic directions ($\gamma = 0$).

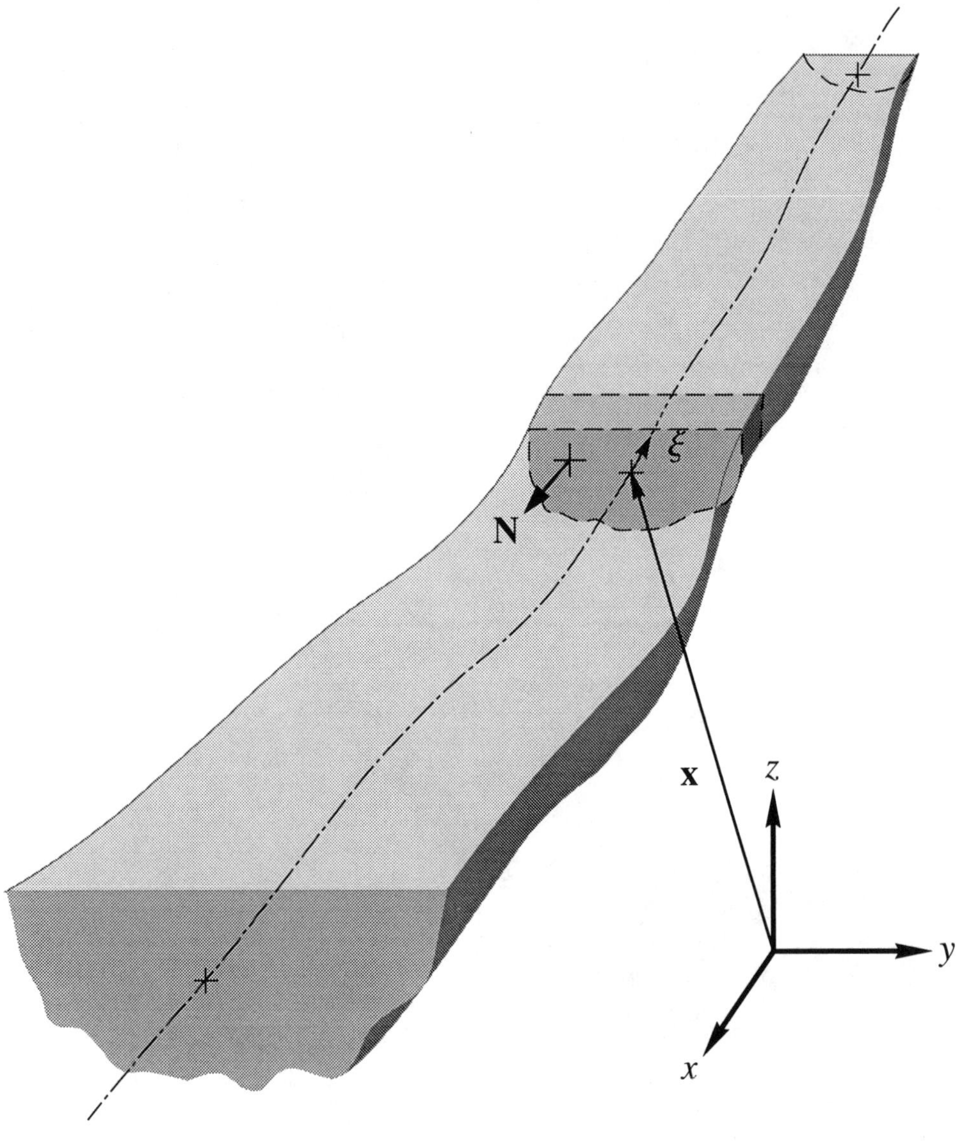

Figure 4.11. Convenient slab REV for megascopic averaging in two dimensions with macroscopic averaging in the third dimension. In contrast to figure 4.10, the extent of the slab REV is only through the region of interest. Thus the spatial generalized function γ is not required to define the megascopic averaging over the cross-section.

spatial and temporal dependence of the limits of integration to the integrand. In this section, the application of generalized functions for this purpose will be demonstrated for simple cases and for cases where both microscopic and macroscopic coordinate systems must be considered. It should be noted that, in general, integration to affect a change in scale may also be performed in the time domain with little additional complication. However, in the current study, the focus will be confined to changes of spatial scales.

4.3.1 General Guidelines

For deriving averaging theorems, two orthogonal coordinate systems are used, $\mathbf{x}$ and $\boldsymbol{\xi}$ (see Section 4.2); $\boldsymbol{\xi}$ is the local coordinate system associated with the averaging region while $\mathbf{x}$ is the global coordinate system. The origin of the $\mathbf{x}$-system is fixed whereas the origin of the $\boldsymbol{\xi}$-system changes with the location of the averaging region in space. For deriving integration theorems, no local coordinate system is employed.

In the most general setting, the generalized functions used to identify curves, surfaces, and volumes of interest, γ^c, γ^s, and γ respectively, are specified as dependent on $\mathbf{x}$, $\boldsymbol{\xi}$, and t. However, in the derivation of integration and averaging theorems, advantage must be taken of the fact that in some instances the spatial dependence is only on $\mathbf{x}$, in other instances only on $\boldsymbol{\xi}$, and in still other instances only on $\mathbf{x}+\boldsymbol{\xi}$. Also, in some cases, the generalized functions do. not depend on time. A number of general guidelines can be established to assist in identification of the proper functional dependence. These guidelines are presented in the context of volumetric generalized functions, but similar considerations apply to generalized functions for surfaces and curves.

1. In the derivation of averaging theorems, if the generalized function is used to identify representative elementary volumes (REV's) that are independent of global position and time, then $\gamma = \gamma(\boldsymbol{\xi})$.
2. If the generalized function is used to identify an entire deforming region used in megascopic averaging (i.e. in deriving integration theorems), then $\gamma = \gamma(\mathbf{x}, t)$. If this region is not deforming, then γ will not depend on t.
3. If the generalized function is used in deriving averaging theorems to identify some portion of all space (not just inside an REV) then γ is a function of spatial position only and not a function of both the location of the REV and the position in the REV relative to its centroid. For this case, $\gamma = \gamma(\mathbf{x}+\boldsymbol{\xi}, t)$.
4. If the generalized function is used in averaging over an REV whose size and/or orientation is spatially and temporally dependent, then $\gamma = \gamma(\mathbf{x}, \boldsymbol{\xi}, t)$.
5. In the derivation of averaging theorems, situations occur where the REV

changes its size or orientation in only one of the coordinate directions. Thus in this direction, dependence of γ on both the global and local coordinate must be accounted for while in the other directions, dependence may be only on the local coordinate system.

The above guidelines are important because the correct identification of the spatial and temporal dependences of the various generalized functions employed in a theorem derivation is critical to the success of the derivation. Perhaps the clearest way to demonstrate proper use of γ, γ^s, and γ^c to delineate regions of interest is through examples as in the following section. These examples are not presented to allow the generalized function to be defined based on the guidelines; rather they serve to clarify the guidelines.

4.3.2 Examples Using Generalized Functions to Delineate Regions

Spherical REV

To identify a spherical volume of radius ρ which is independent of time and its location in space (see figure 4.7) then, according to Guideline 1, $\gamma = \gamma(\boldsymbol{\xi})$ where:

$$\gamma(\boldsymbol{\xi}) = \begin{cases} 1 \text{ when } |\boldsymbol{\xi}| < \rho \\ 0 \text{ when } |\boldsymbol{\xi}| > \rho \end{cases} \tag{4.2}$$

If the size of the sphere depends on time but not the position of the sphere, then:

$$\gamma(\boldsymbol{\xi}, t) = \begin{cases} 1 \text{ when } |\boldsymbol{\xi}| < \rho(t) \\ 0 \text{ when } |\boldsymbol{\xi}| > \rho(t) \end{cases} \tag{4.3}$$

The region described by equation (4.2) is particularly useful as an REV for macroscopic averaging. By integrating a property over the $\boldsymbol{\xi}$ coordinate, assigning the average value of the property to the centroid of the REV, and then repeating the process for the REV located at all points in the megascopic region under study, one obtains a continuous function for the averaged property.

Spatially Dependent REV

Consider the case when γ is used to indicate an averaging volume whose size may be a function of position but not time. This situation may arise in hydrology for an REV used to average over a heterogeneous porous medium where the averaging volume increases as the average grain size increases. In this case, $\gamma = \gamma(\mathbf{x}, \boldsymbol{\xi})$ because the value of γ depends independently on $\mathbf{x}$ and $\boldsymbol{\xi}$ (Guideline 4).

Suppose, for illustration, that with increasing distance from the origin of the $\mathbf{x}$-coordinate system, a larger averaging volume is used so that:

$$\gamma(\mathbf{x}, \xi) = \begin{cases} 1 \text{ when } |\xi| < \rho(\mathbf{x}) \\ 0 \text{ when } |\xi| > \rho(\mathbf{x}) \end{cases} \qquad \textbf{(4.4)}$$

where $\rho(\mathbf{x})$ is an increasing function of $|\mathbf{x}|$. Now consider figure 4.12. If $\xi_1 = \xi_2 = \xi$, then $\gamma = 0$ for $\mathbf{x}_1 + \xi$ but $\gamma = 1$ for $\mathbf{x}_2 + \xi$. Thus γ is not a function of ξ only (as would be the case for a constant-radius REV); but γ must depend on $\mathbf{x}$ and ξ independently, i.e. $\gamma = \gamma(\mathbf{x}, \xi)$. However, depending on the function $\rho(\mathbf{x})$, functional dependence on some combination of $\mathbf{x}$ and ξ other than $\mathbf{x} + \xi$ may be found. If the size of the REV changes with time also, then $\gamma = \gamma(\mathbf{x}, \xi, t)$.

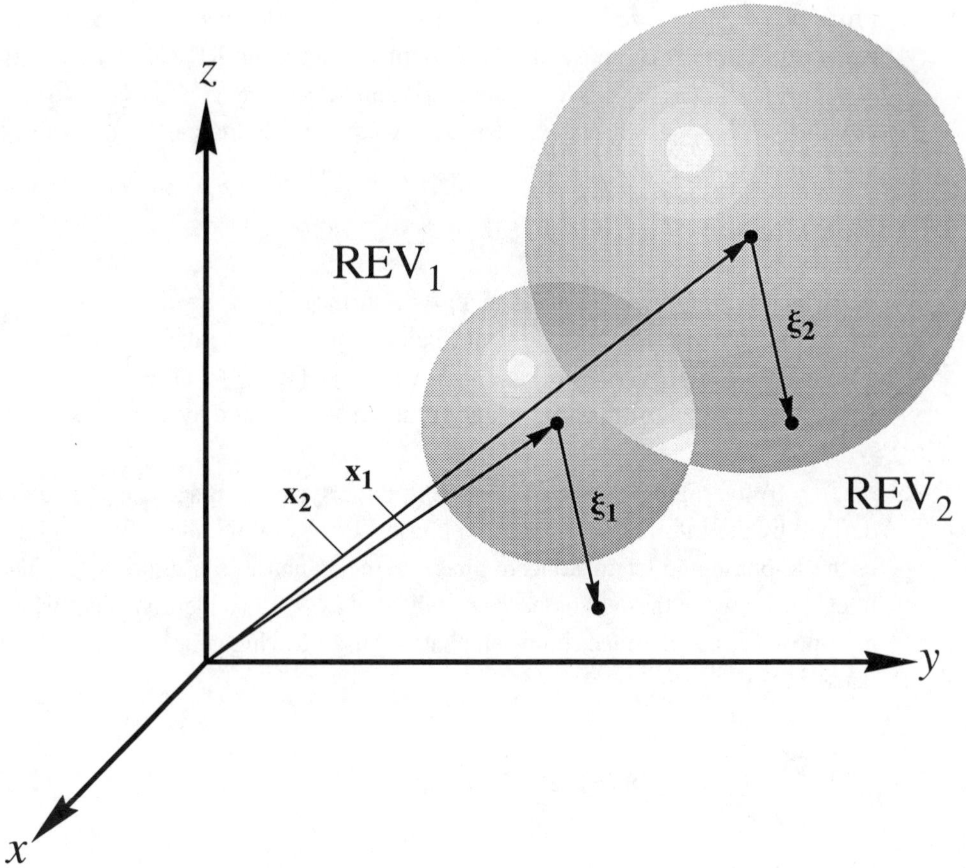

Figure 4.12. Spatially dependent REVs for which γ is a function of $\mathbf{x}$ and ξ rather than $\mathbf{x}+\xi$ as with a a constant REV.

Location of an α-Phase in All Space

Suppose that the region occupied by one phase, referred to as the α-phase, in a multiphase system is to be delineated. For this case, the value of the generalized function depends on position in the domain, such that if a point is in the α-phase, $\gamma = 1$; otherwise $\gamma = 0$. Since the position of the α-phase in general changes with time, and given that a position $\mathbf{x} + \boldsymbol{\xi}$ is unique, then $\gamma^{\alpha} = \gamma^{\alpha}(\mathbf{x} + \boldsymbol{\xi}, t)$ where the superscript α is used to indicate that this generalized function identifies the α-phase in all space. This representation of functional dependence is consistent with Guideline 3. Note that the relation to the physical system under study, as opposed to identification of a region for mathematical purposes, is a significant conceptual difference between γ^{α} and the generalized function, γ, used in the previous examples to define an REV.

Location of an α-Phase Inside an REV

Products of generalized functions can be used to identify more specialized regions of interest. Suppose all of the α-phase inside an REV is to be identified. If $\gamma = \gamma(\boldsymbol{\xi})$ defines a constant REV, and if $\gamma^{\alpha} = \gamma^{\alpha}(\mathbf{x} + \boldsymbol{\xi}, t)$ identifies the α-phase in all space, then the product $\gamma\gamma^{\alpha}$ indicates all α-phase portions residing inside the REV.

Identification of a General Surface

A surface is identified in space using the surface generalized function, γ^{s}, which depends on two surficial coordinates. For use in the derivation of integration theorems over a surface, $\gamma^{s}(\mathbf{u}) = \gamma^{s}[\mathbf{u}(\mathbf{x})]$, as elucidated by Guideline 2. The portion of the surface of interest is located by all points where $\gamma^{s} = 1$.

In averaging theorems for multiphase systems, a surface is usually defined by the interface between two phases. For example, identify one phase as the α-phase and let an adjacent phase be the β-phase (see figure 4.13). The interface between these two phases is indicated as $S_{\alpha\beta}$. To identify this surface as opposed to all interfaces between phases, the following generalized function is used:

$$\gamma^{s} = \gamma^{s}[\mathbf{u}(\mathbf{x} + \boldsymbol{\xi}, t)] = \begin{cases} 1 \text{ when } & (\mathbf{u} \in S_{\alpha\beta}) \\ 0 \text{ when } & (\mathbf{u} \notin S_{\alpha\beta}) \end{cases} \tag{4.5}$$

Note that $S_{\alpha\beta}$ could be a finite number of non-overlapping simple regions. The functional dependence of γ^{s} is consistent with Guideline 3.

As with the previous example, products of generalized functions can be used to restrict $S_{\alpha\beta}$ to an REV. For example consider $\gamma\gamma^{s}$ where γ is

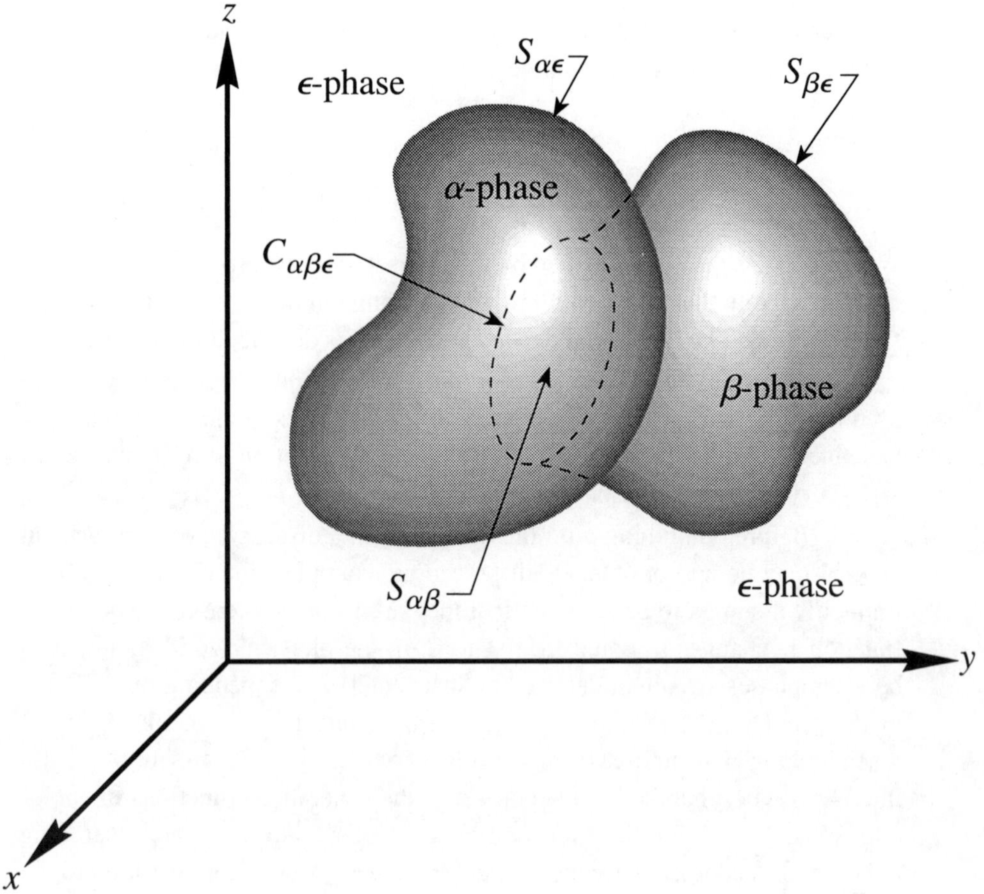

Figure 4.13. Interface $S_{\alpha\beta}$ between adjacent α and β phases. This interface may be distinguished from other phase interfaces using the surficial generalized function γ^s defined as in equation (4.5).

defined by equation (4.2) and γ^s is defined by equation (4.5). This product identifies all $\alpha\beta$-interfaces inside an REV whose shape and orientation are independent of position and time.

Identification of a General Curve

A space curve is identified using the generalized function, γ^c, which depends on the curvilineal coordinate system $\mathbf{l}$ that has the single curvilineal linear coordinate l. For use in derivation of integration theorems for a general simple curve, $\gamma^c = \gamma^c[\mathbf{l}(\mathbf{x}), t]$ by Guideline 2. The portion of the curve of interest is located by all points where $\gamma^c = 1$.

In averaging theorems for multiphase systems, a curve is formed by the intersection of three distinct phases (see figure 4.13). To identify this curve using generalized functions, define:

$$\gamma^c = \gamma^c[\mathbf{l}(\mathbf{x}+\boldsymbol{\xi}),t] = \begin{cases} 1 \text{ when } (\mathbf{l} \in C_{\alpha\beta\epsilon}) \\ 0 \text{ when } (\mathbf{l} \notin C_{\alpha\beta\epsilon}) \end{cases} \quad \textbf{(4.6)}$$

Where $C_{\alpha\beta\epsilon}$ is the curve formed by the intersection of the $\alpha\beta$-, $\beta\epsilon$-, and $\alpha\epsilon$-interfaces. Note that $C_{\alpha\beta\epsilon}$ could represent a finite number of non-intersecting simple curves. The functional dependence of γ^c is described in Guideline 3.

As in previous examples, products of generalized functions can be used to restrict $C_{\alpha\beta\epsilon}$ to certain regions in space. For example, $\gamma\gamma^c$, where γ is defined by equation (4.2) and γ^c is defined by equation (4.6), identifies all $\alpha\beta\epsilon$-curves inside a constant-sized REV.

Before continuing with the next section, a discussion of interfaces in general may be appropriate. In all preceding examples, interfaces have been implicitly assumed to be sharp so that they are located where the generalized function γ changes in value from 1 to 0. In the physical world, boundaries between phases are often idealized as sharp interfaces depending on the scale of the problem. An example would be a boundary between geologic layers when looking at an aquifer from a regional point of view. Treatment of diffuse interfaces is best handled in the context of the generalized functions in one of two ways: 1) defining the transition zone as a separate phase and averaging over that phase, or 2) assigning properties such as mass and momentum to the interface. For the former, refer to *Hassanizadeh and Gray* [1989] while the latter is presented by *Gray and Hassanizadeh* [1989].

4.4 INTERCHANGE OF INTEGRATION REGIONS

In the development of integration and averaging theorems for curves and surfaces, the ability to transform integrals over curves or surfaces to volume integrals is a very useful skill. The reason for this can be seen by considering one of the cases, for example a surface. By introduction of the generalized function γ^s, an integral over a finite simple surface can be transformed to an integral over S_∞, i.e. an infinite extension of the surface of interest. At this point, however, time and/or space derivatives cannot be moved outside the integral sign because S_∞ may still depend on time or space. A surface - even S_∞ - is two-dimensional so it can change its location and shape in three-dimensional space. However, by transforming the surface integral to a volume integral over all space, i.e. V_∞, the integration limits no longer show spatial or temporal dependence. This transformation, from a surface to a volume integral, is achieved through the use of generalized functions. In essence, the integral properties presented as equations (3.39), (3.40a), and (3.40b) are used to change from a surface integration to a volume integration. Similarly integrations over spatial curves may be transformed to integrals over surfaces using equations (3.32), (3.33a), or (3.33b); and the resulting surface integrals may be further transformed to integrals over volumes.

The manner in which curves and a surfaces are extended for purposes of integration is not arbitrary. All unit normal vectors normal to the surfaces or volumes which result must be orientable. Recall that an orientable vector is defined by a continuous function and has a unique direction associated with it for each point where it exists. Thus, curves and surfaces which define the boundary of a region must be simple. Surfaces cannot fold back over themselves or be twisted as in a Möbius strip. Curves are not allowed to intersect themselves. These restrictions are imposed to ensure that the outward normal direction to a surface or volume can be identified. They do not provide particularly severe limitations on the applicability of generalized functions to the development of integration theorems in that complex surfaces and curves may usually be made orientable by breaking them into pieces and considering each piece separately.

In converting integration regions, each gradient of a generalized function that appears in an integrand serves to reduce the dimension of an integration problem by one. Specifically when a gradient of a generalized function appears in the integrand, a volume integral becomes a surface integral; a surface integral becomes an integral along a curve; a curvilineal integral becomes point evaluations with all evaluations occurring at points where the derivative of the generalized function is non-zero. Because the facilitation of interchange among volume, surface, and curvilineal integrals is an extremely powerful property of the generalized functions employed in this work, this feature is examined in additional detail in the following subsections.

4.4.1 Transformation of Integration over a Surface to Integration over a Volume

Consider an orientable surface S that is a portion of a closed simple surface denoted as S_∞ (see figure 4.14). Then using the surficial generalized function, one may convert an integral over S to an integral over S_∞ as:

$$\int_S f dS = \int_{S_\infty} f\gamma^S dS \tag{4.7}$$

Because S_∞ is a closed surface, it may be treated as the boundary of a volume V. A spatial generalized function, γ, may now be defined to identify this volume using the convention of equation (3.36). Then by equation (3.39), where $\mathbf{n}$ is the outwardly directed unit vector normal to the volume:

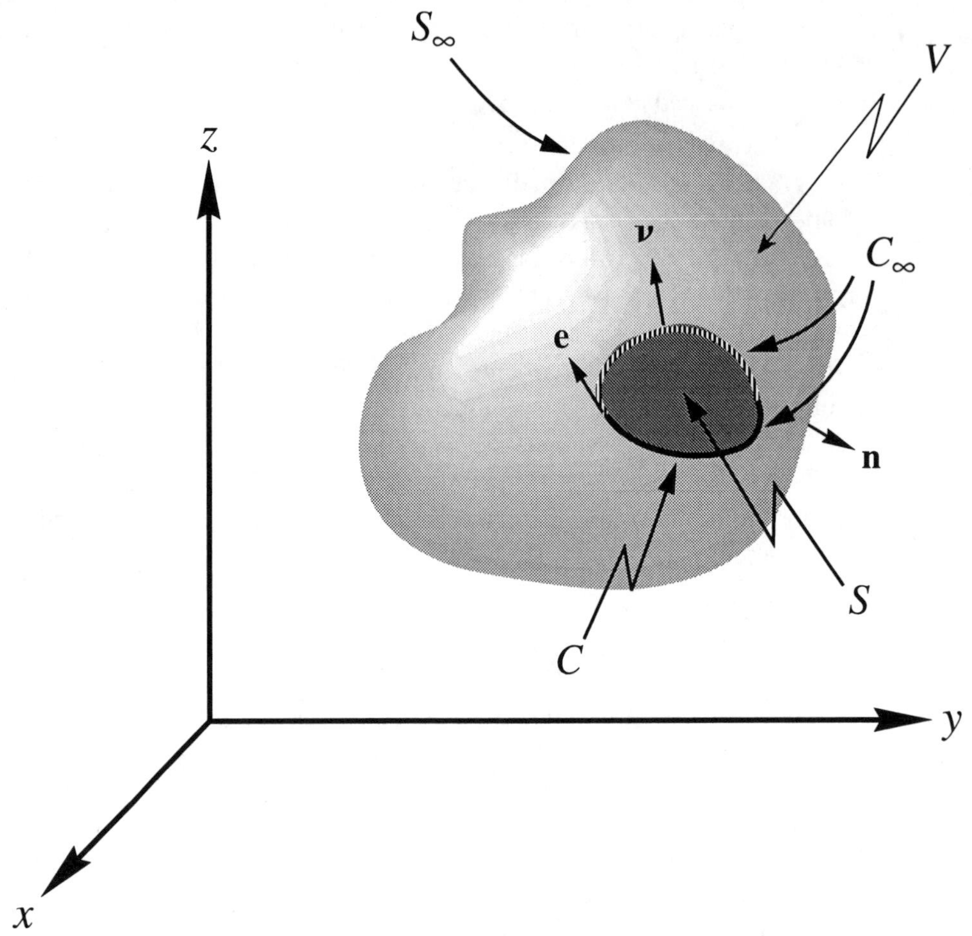

Figure 4.14. Surface S that is a portion of a simple closed surface, S_∞, that bounds volume V. The boundary of S is a closed curve C_∞. The curve C is a portion of C_∞. Unit vectors are defined such that $\mathbf{n}$ is normal to S, $\boldsymbol{\nu}$ is normal to C_∞ and tangent to S_∞, and $\mathbf{e}$ is tangent to C and tangent to S_∞.

$$\int_S f dS = -\int_{V_\infty} f \gamma^s \mathbf{n} \cdot \nabla \gamma dV \tag{4.8}$$

It is important in writing relations such as equation (4.8) to maintain the distinction between surficial and spatial generalized functions and to correctly identify the unit vector.

4.4.2 Transformation of Integration over a Curve to Integration over a Surface

Suppose integration of some function f is to be carried out over some finite simple curve C. This curve may be extended to form a simple closed loop which is identified, for convenience, as C_∞. An integral over C can be written as an integral over C_∞ by making use of the curvilineal generalized function γ^c defined to be equal to 1 over C and 0 over the rest of C_∞. Thus:

$$\int_C f dC = \int_{C_\infty} f\gamma^c dC \tag{4.9}$$

Next consider the closed loop defined by C_∞ to be the curve bounding a simple orientable surface S. The unit normal vector to C_∞ which is tangent to and pointing outward from S is $\boldsymbol{\nu}$. Also let S_∞ be a closed surface which contains S. If γ^s is defined to equal 1 on S but zero on the remainder of S_∞, equation (3.32) may be invoked to transform the right side of equation (4.9) to a surface integral such that:

$$\int_C f dC = -\int_{S_\infty} f\gamma^c (\boldsymbol{\nu}\bullet\nabla^s\gamma^s)\, dS \tag{4.10}$$

At this point the closed surface, S_∞, can be considered to be the boundary of a volume V. This volume is a subregion of all space, V_∞. Let γ be the generalized function locating volume V such that γ equals 1 in V but is 0 in the rest of V_∞. If the outwardly directed unit normal to V is identified as $\mathbf{n}$, equation (3.39) can be invoked to change the right side of equation (4.10) to an integral over V_∞:

$$\int_C f dC = \int_{V_\infty} f\gamma^c (\boldsymbol{\nu}\bullet\nabla^s\gamma^s)\,(\mathbf{n}\bullet\nabla\gamma)\, dV \tag{4.11}$$

Equations (4.9) through (4.11) provide the successive steps for changing an integral over a curve to an integral over a surface, and the integral over that surface to an integral over a volume. It is important to have an understanding of these manipulations and to be able to relate them to a physical system under study. For example, if the integration over the interface $S_{\alpha\beta}$ between the α and β phases within an REV in a multiphase system is of interest, generalized functions can be used to change the integral to one over V_∞. Define the following three generalized functions:

$\gamma(\boldsymbol{\xi})$ is a spatial generalized function for an REV as described in equation (4.2).

$\gamma^{\alpha}(\mathbf{x}+\boldsymbol{\xi}, t)$ is the spatial generalized function that locates the α-phase in space as described in the third example of Section 4.3.2.

$\gamma^{s}[\mathbf{u}(\mathbf{x}+\boldsymbol{\xi}), t]$ is the surficial generalized function that identifies $S_{\alpha\beta}$.

If $f(\mathbf{x}+\boldsymbol{\xi}, t)$ is the function being integrated over $S_{\alpha\beta}$, then:

$$\int_{S_{\alpha\beta}} f dS = -\int_{V_{\infty}} f\gamma^{s}\gamma(-\mathbf{n}\bullet\nabla\gamma^{\alpha})\, dV \tag{4.12}$$

One other important relation between integrals over a surface and a curve is obtained by considering the integral over a surface S of the surface gradient. If integration is to be extended from being over S to integration over a closed surface, S_{∞}, that contains S, a generalized function must be introduced. The surficial generalized function may be introduced such that:

$$\int_{S} \nabla^{s} f dS = \int_{S_{\infty}} (\nabla^{s} f)\, \gamma^{s} dS \tag{4.13}$$

An alternative approach to facilitating the transformation from integration over a finite surface to integration over S_{∞} is to consider the surface of integration as being one that intersects a volume of interest as in figure 4.15. Then the generalized function for the volume may be used to identify the portion of S_{∞} intersecting that volume to form the surface of interest. Thus an alternative expression to equation (4.13) is:

$$\int_{S} \nabla^{s} f dS = \int_{S_{\infty}} (\nabla^{s} f)\, \gamma dS \tag{4.14}$$

Although the difference between these expressions may appear to be minor, in fact some important information may be obtained. The normal to the surface will be denoted as $\mathbf{n}$ while the normal to the edge of the surface that is also tangent to the surface will be denoted as $\boldsymbol{\nu}$. These two vectors are orthogonal at the edge of the surface. The normal to the volume at the edge of the surface is denoted as $\mathbf{n}^{*}$ and is equal to $\boldsymbol{\nu}$ only if the surface of interest is orthogonal to the boundary of the volume. The right sides of equations (4.13) and (4.14) are equal. Thus integration by parts may be employed while preserving the equality to obtain:

$$\int_{S_{\infty}} \nabla^{s}(f\gamma^{s})\, dS - \int_{S_{\infty}} f\nabla^{s}\gamma^{s} dS = \int_{S_{\infty}} \nabla^{s}(f\gamma)\, dS - \int_{S_{\infty}} f\nabla^{s}\gamma\, dS \tag{4.15}$$

The first terms on both sides of this equation are zero (This can be readily

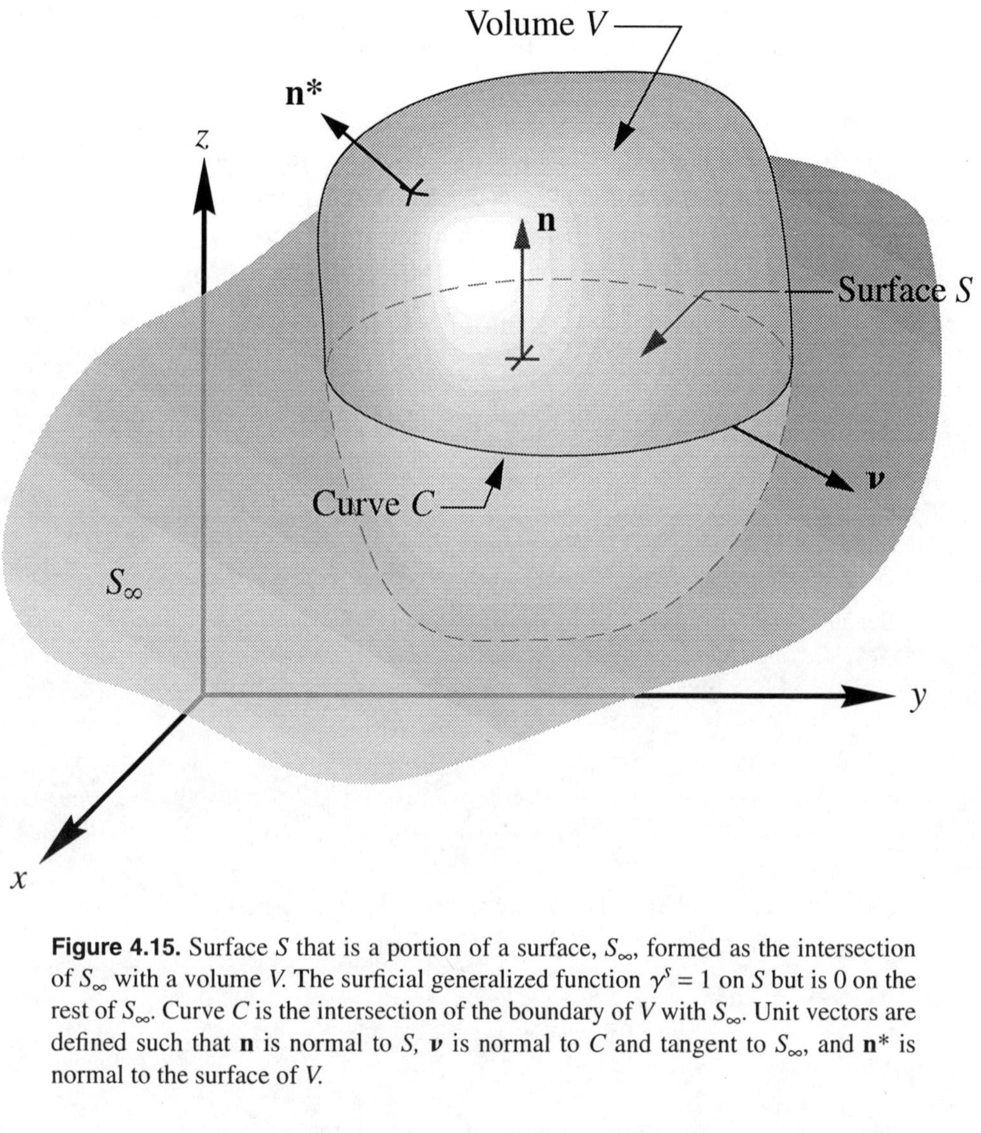

Figure 4.15. Surface S that is a portion of a surface, S_∞, formed as the intersection of S_∞ with a volume V. The surficial generalized function $\gamma^s = 1$ on S but is 0 on the rest of S_∞. Curve C is the intersection of the boundary of V with S_∞. Unit vectors are defined such that $\mathbf{n}$ is normal to S, $\boldsymbol{\nu}$ is normal to C and tangent to S_∞, and $\mathbf{n}^*$ is normal to the surface of V.

shown by a series of manipulations related to the theorem derivation procedure presented in Chapter 6. The steps are not presented here in order to keep focus on the point of this section.). Equation (2.26) relates a surface gradient to a spatial gradient and may be employed to express equation (4.15) as:

$$\int_{S_\infty} f\nabla^s \gamma^s dS = \int_{S_\infty} f\nabla\gamma\, dS - \int_{S_\infty} f\mathbf{nn}\cdot\nabla\gamma dS \qquad \textbf{(4.16)}$$

The left side of this equation is readily converted to an integral over the closed curve bounding S, designated as C, using equation (3.33b). Now note that $\nabla\gamma$, which is non-zero only on the boundary of the volume, will be non-zero only where S_∞ and the boundary of the volume intersect, i.e. only on C. However, because $\nabla\gamma$ is the spatial derivative of a spatial generalized function, the magnitude of its integral, though appropriate for conversion of a volume integral to a surface integral, is not known when converting a surface integral to an integral over a curve. For convenience, denote this magnitude as $\mathcal{M}$ such that the integrals in equation (4.16) convert to integrals over a curve as follows:

$$-\int_C f\boldsymbol{\nu}\,dC = -\int_C f\mathcal{M}\mathbf{n}^*\,dC + \int_C f\mathcal{M}\mathbf{n}\mathbf{n}\cdot\mathbf{n}^*\,dC \tag{4.17}$$

Because the integration here is not dependent on any particular curve geometry, the expression may be localized indicating that at any point on C, the equality must hold such that:

$$\boldsymbol{\nu} = \mathcal{M}\mathbf{n}^* - \mathcal{M}\mathbf{n}\mathbf{n}\cdot\mathbf{n}^* \tag{4.18}$$

Because $\boldsymbol{\nu}$ is orthogonal to $\mathbf{n}$, the dot product of $\boldsymbol{\nu}$ with this equation yields:

$$1 = \mathcal{M}\boldsymbol{\nu}\cdot\mathbf{n}^* \tag{4.19a}$$

or:

$$\mathcal{M} = \frac{1}{(\boldsymbol{\nu}\cdot\mathbf{n}^*)} \tag{4.19b}$$

From this expression, and equating the corresponding first terms on the right sides of equations (4.16) and (4.17), one obtains:

$$\int_{S_\infty} f\nabla\gamma\,dS = -\int_C \frac{f\mathbf{n}^*}{(\boldsymbol{\nu}\cdot\mathbf{n}^*)}dC \tag{4.20}$$

This equation will prove to be particularly helpful in the derivation of integration theorems when integration is over a planar surface. Although C in equation (4.20) is a closed simple curve, the inclusion of the curvilineal generalized function γ^c in the integrand on both sides of the equation allows it to apply when integration is over a portion of a closed curve. Note also that the surface integral on the left side of equation (4.20) may be converted to a volume integral as discussed in Subsection 4.4.1.

4.4.3 Transformation of Integration over a Curve to Evaluation at the Ends of the Curve

This kind of transformation is needed when integrating a curvilineal del operator applied to a function over a simple curve. For example, if curve C in figure

4.14 is the part of a closed curve C_∞ where $\gamma^c = 1$:

$$\int_C \nabla^c f dC = \int_{C_\infty} (\nabla^c f)\, \gamma^c dC \tag{4.21}$$

Application of the chain rule to the right side of this equation and noting that integration of $\nabla^c (f\gamma^c)$ along a closed curve will be zero rearranges this equation to the form:

$$\int_C \nabla^c f dC = -\int_{C_\infty} f \nabla^c \gamma^c dC \tag{4.22}$$

Then from equation (3.26b), the right side of this equation is converted to an evaluation of the integrand at the ends of C such that:

$$\int_C \nabla^c f dC = (\mathbf{e}f)\big|_{\text{ends}} \tag{4.23}$$

where $\mathbf{e}$ is a vector tangent to the curve and pointing outward from C at each end. If C is a closed curve such that it has no "ends," the right side of equation (4.23) will be zero.

When integration over a curve is being considered, the generalized function for a surface or a volume may be used to identify a portion of the curve as lying in a surface or intersecting a volume. The development of the relation between the integration over the curve and the evaluation at the end points is analogous to that presented in the last subsection for a surface intersecting a volume. Only the results will be presented here. For a curve C contained in a surface S as in figure 4.16:

$$\int_{C_\infty} f \nabla^s \gamma^s dC = \left.\frac{-f\boldsymbol{\nu}^*}{(\mathbf{e}\cdot\boldsymbol{\nu}^*)}\right|_{\text{ends}} \tag{4.24}$$

where $\boldsymbol{\nu}^*$ is the outward directed normal to the boundary of the surface that is tangent to the surface at the intersection points of C and the boundary of the surface. For a curve C intersecting a volume V as in figure 4.17:

$$\int_{C_\infty} f \nabla \gamma dC = \left.\frac{-f\mathbf{n}^*}{(\mathbf{e}\cdot\mathbf{n}^*)}\right|_{\text{ends}} \tag{4.25}$$

where $\mathbf{n}^*$ is the outwardly directed unit normal to the surface of the volume at the points where the curve intersects the surface.

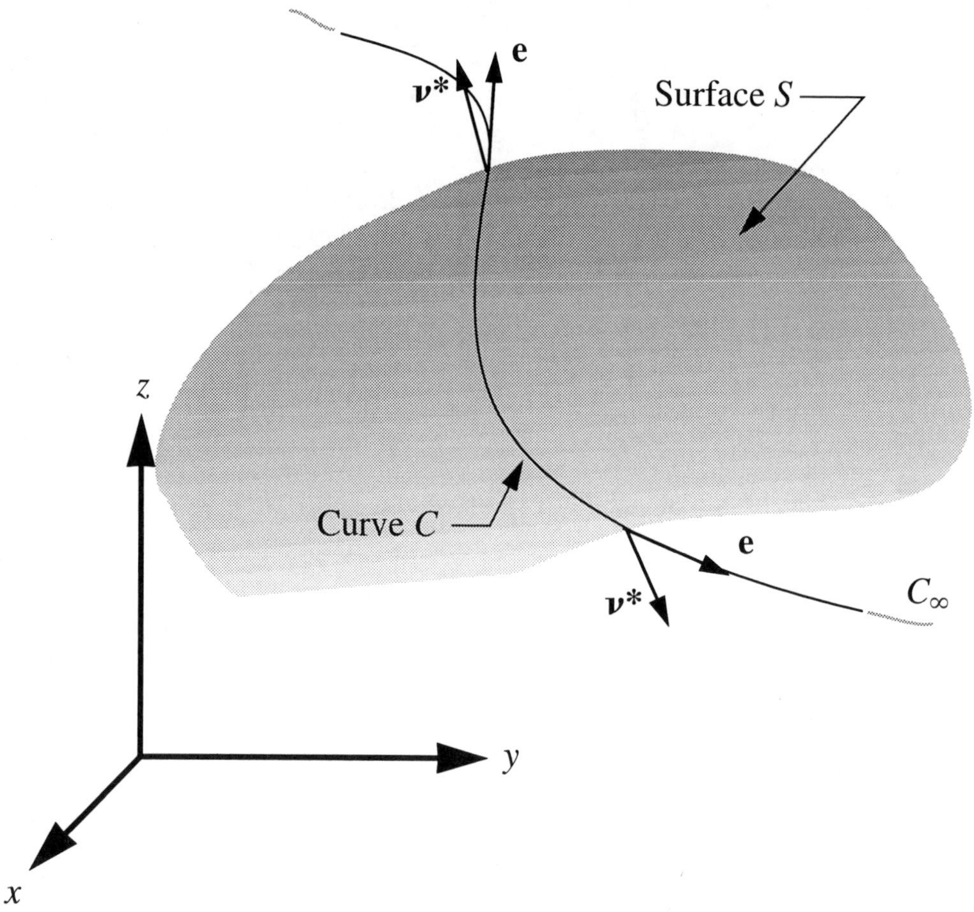

Figure 4.16. Curve C, a portion of a curve, C_∞, lying in a surface S. The surficial generalized function $\gamma^S = 1$ on S but is 0 in the rest of space. The endpoints of curve C are the intersection of the boundary of S with C_∞. Unit vectors are defined such that $\mathbf{e}$ is tangent to C pointing outward from S and $\boldsymbol{\nu}^*$ is normal to the boundary of S and tangent to S.

4.4.4 Summary of the Gradients of the Generalized Functions Used in this Work

An important concept in the preceding sections is the dimensionality of the generalized function. This will be summarized for emphasis and clarity. For curves, $\boldsymbol{\lambda} \cdot \nabla^C \gamma^C$ is non-zero at the end points of the indicated region where $\boldsymbol{\lambda}$ is the tangent to the curve. For surfaces, $\boldsymbol{\nu} \cdot \nabla^S \gamma^S$ is non-zero along a curve bounding a region on the surface and $\boldsymbol{\nu}$ is a unit vector tangent to the surface

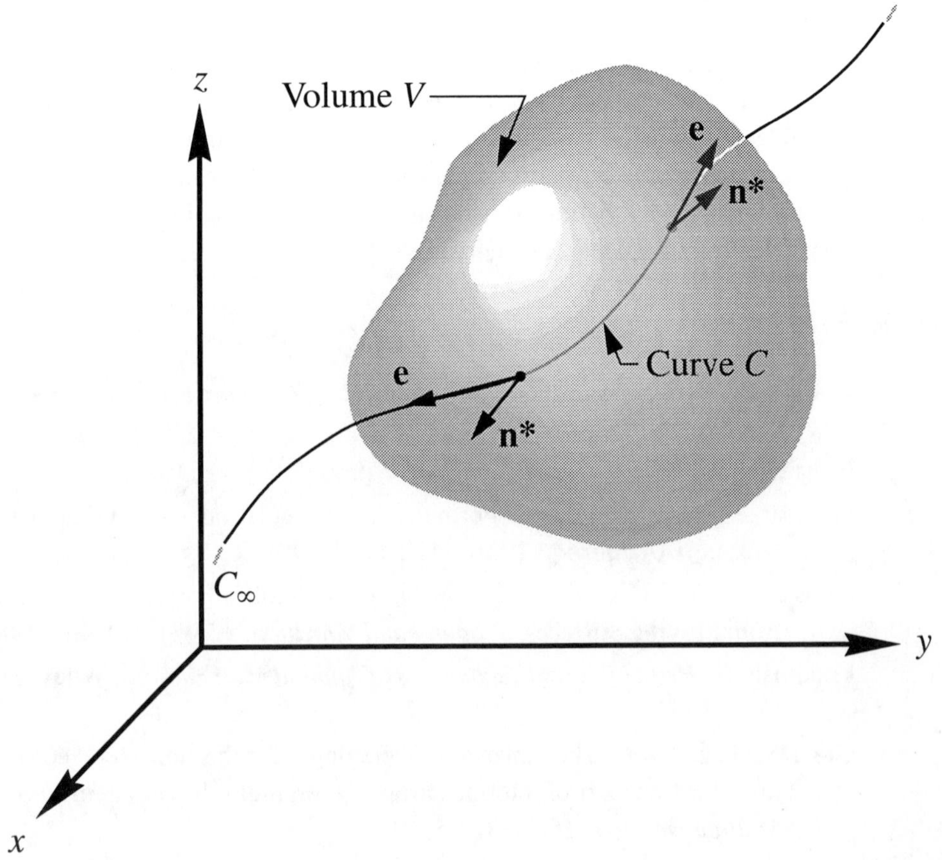

Figure 4.17. Curve C, a portion of a curve, C_∞, intersecting a volume V. The spatial generalized function $\gamma = 1$ in V but is 0 in the rest of space. The endpoints of curve C are the intersection of the boundary of V with C_∞. Unit vectors are defined such that **e** is tangent to C pointing outward from V and **n*** is normal to the boundary of V.

and normal to the bounding curve. For volumes, $\mathbf{n} \cdot \nabla \gamma$ is non-zero along the surface that bounds the volume in space and **n** is the normal to the volume. In the latter two cases, the gradient of the generalized function is non-zero at an infinite number of points. This generalization is not unique to this work. For example, *Farassat* [1981] and *Farassat and Myers* [1988] have applied generalized functions in the field of acoustics; *Kanwal* [1983] applies generalized functions to solution of ordinary and partial differential equations; and *Friedman* [1956] notes that physicists use this type of distribution to represent line and surface charges.

4.5 CONCLUSION

The generalized function has been shown to be a very useful tool for identification of portions of a curve, a surface, or space. Exploitation of this property facilitates the transformation of an integral over some subregion to integration over a global region. Furthermore, the properties of derivatives of generalized functions make the functions valuable tools for transformations among integrations over volumes, areas, and curves.

4.6 REFERENCES

Bird, R. B, W. E. Stewart, and E. N. Lightfoot, *Transport Phenomena*, Wiley, New York, 1960.

Eringen, A. C., *Mechanics of Continua*, Wiley, New York, 1967.

Farassat, F., Linear Acoustic Formulas for Calculation of Rotating Blade Noise, *AIAA Journal*, **19**(9), 1122-1130, 1981.

Farassat, F., and M. K. Myers, Extension of Kirchhoff's Formula to Radiation from Moving Surfaces, *J. Sound and Vibration*, **123**(3), 451-460, 1988.

Friedman, B., *Principles and Techniques of Applied Mathematics*, Wiley, New York, 1956.

Gray, W. G., and S. M. Hassanizadeh, Averaging theorems and averaged equations for transport of interface properties in multiphase systems, *Intl. J. Multiphase Flow*, **15**(1), 81-95, 1989.

Hassanizadeh, S. M., and W. G. Gray, Derivation of conditions describing transport across zones of reduced dynamics within multiphase systems, *Water Resources Research*, **25**(3), 529-539, 1989.

Kanwal, R. P., *Generalized Functions - Theory and Technique*, Academic Press, New York, 1983.

Malvern, L. E., *Introduction to the Mechanics of a Continuous Medium*, Prentice-Hall, Englewood Cliffs, 1969.

Whitaker, S., Advances in theory of fluid motion in porous media, *Industrial and Engineering Chemistry*, **61**(12), 14-28, 1969.

CHAPTER FIVE

THEOREM NOTATION

5.0 INTRODUCTION

In order to facilitate the development and cataloguing of the variety of theorems describing the transformation of derivatives from one scale to another, a consistent and convenient notational scheme is employed. Within this nomenclature scheme, two important classes of theorems are of particular interest: 1) integration theorems that involve only microscopic and megascopic scales, and 2) averaging theorems that always involve microscopic and macroscopic scales but may also involve the megascopic scale. The starting point for both classes of theorems is the integration of the derivative of a function describing microscale behavior. Through this integration, the function is transformed to either the macroscale, the megascale, or some mix of the two scales. Furthermore, if the transformation is to the macroscale, the spatial derivatives are also transformed to that scale.

The notation scheme used to identify the 128 theorems that result is useful both as a method of categorizing the theorems and for location of a theorem affecting a particular transformation in Chapters 7 and 8. Thus, although the current chapter is concerned only with an exposition of the notation used, mastery of the notational scheme increases the accessibility of the subsequent theorem tabulations.

5.1 GENERAL THEOREM NOTATION

Integration theorems convert a gradient, divergence, curl, or partial time derivative of a function at the microscale to those observed from a megascopic perspective. This transformation may be applied in one, two, or three spatial dimensions with microscopic variation being retained in the other dimensions. Through this conversion, a function becomes either totally megascopic or some mix of microscopically varying and megascopic. Information concerning microscopic variation along any megascopic direction is lost. In contrast, averaging theorems transform a microscale gradient, divergence, curl, or time derivative of a function to functions and derivatives observed using some mix of the macro- and megascales. With averaging theorems, all explicit knowledge of microscale variation is lost.

The general notation used to identify one of these theorems is:

$$\langle \text{letter} \rangle \, [i, \, (m,n) \, , k]$$

where:

- $\langle$letter$\rangle$ will be D, G, C, or T to identify the theorem as dealing with the divergence, gradient, curl or time derivative, respectively (Theorems dealing with the time derivative are also referred to as transport theorems.),
- i indicates the microscopic dimensionality of the del or time derivative operator in the theorem,
- m and n relate to the macroscopic dimensionality of the theorem, and
- k indicates the megascopic dimensionality of the theorem.

The indices in this notation are significant for identification of the type of transformation occurring. Thus knowledge of the type of scale transformation that one wishes to employ in a particular application and of the theorem identification notation facilitates easy location of the appropriate theorem. Thus, a fuller discussion of these indices follows.

The first index, i, is related to the microscopic dimensionality of the operator undergoing transformation. For a spatial operator, $i = 3$; for a surficial operator, $i = 2$; and for an operator defined along a curve, a curvilineal operator, $i = 1$. For the T (transport or time derivative) theorems, i is the number of position coordinates held constant when the partial time derivative is evaluated. For example, when a partial time derivative is evaluated at fixed coordinates on a surface, $i = 2$, while for a partial time derivative evaluated at a fixed position in space, $i = 3$. The theorems will be presented with the operators acting on spatial functions (or vectors) with the simplification to functions of lower dimensionality being straightforward and only requiring that certain terms in the theorems be set equal to zero.

The sum of the two indices in parentheses in the theorem identification notation, $m+n$, is the number of macroscopic dimensions appearing in the resulting theorem. In the most general case, $0 \le m$, $0 \le n$, and $0 \le m+n \le 3$. However, in the current text, only a subset consisting of the most useful theorems, rather than all possible combinations of m and n, is considered. When $m+n = 0$, the theorem is an integration theorem. These theorems involve transformation of some or all of the dependence on spatial dimensions from the microscale to the megascale and appear in Chapter 7. The averaging theorems considered are those for which $1 \leq m+n \leq 3$ subject to the additional constraint that $m \times n = 0$. It is important to note that averaging theorems make use of integration over space (when $i = 3$), surfaces (when $i = 2$), or curves (when $i = 1$) within an REV. When m is the non-zero index, the transformation from the microscale obtained by integration over an REV results in terms that include macroscopic spatial derivatives in m spatial coordinates. When n is the non-zero index, the form of the transformation is written with flux terms at the boundary of the REV instead of spatial derivative terms. In both cases, the resulting expression is dependent on $m+n$ macroscopic dimensions. The averaging theorems are tabulated in Chapter 8.

The fourth index employed in the theorem identification convention, k, provides the megascopic dimensionality. For integration theorems, this index thus indicates how many of the microscopic coordinates have been integrated to the megascale. For these theorems, when $k < 3$, some of the coordinates may remain at the microscale. On the other hand, the derivation of the averaging theorems makes use of integration over i-dimensional space within an REV so that $m+n+k = 3$, where k is the number of megascopic dimensions while $m+n$ is the number of macroscopic dimensions. Note that with the averaging theorems, a microscopic function that may depend on two surficial coordinates (or one curvilineal coordinate) is averaged over the surface (or curve) contained within an REV positioned in space. The result is that the averaged quantities are associated with each position of the REV and the sum of macroscale plus megascale dimensions must be 3.

As mentioned above, only those averaging theorems whose product $m \times n$ is identically zero are discussed and presented here. Theorems whose product $m \times n$ is not zero involve terms with implicit macroscale dependence in n dimensions in addition to macroscopic derivative terms in m dimensions. For example, a theorem with the indices $[3, (2, 1), 0]$ would have derivatives at the macroscale in two dimensions ($m = 2$) as well as dependence on the third macroscale dimension through integrals over the surface of the REV normal to that third dimension. Theorems of this type would seem to have dubious utility in practice and hence are excluded from this study. In fact, the averaging theorems for which $m > 0$ and $n = 0$ would seem to be the most useful forms in applications.

The purpose of the next section of this chapter is to provide insight into the relationship between the theorem identification convention and the forms of terms which appear in the theorems. This chapter is not concerned with illustrating the development of the theorems or with their utility. The goal here is to develop facility in identifying the structure of the theorems by using the classification scheme.

5.2 NOTATION SPECIFIC TO THEOREM CLASSIFICATIONS

5.2.1 Integration Theorems

For this work, integration theorems are those theorems which relate integrals over some region of a spatial, surficial, or curvilineal derivative of a function to integrals over the boundary of that region. Additionally, integration theorems transform the integral over a region of a time derivative of a function to a combination of a time derivative of the integral plus integrals over the boundary of the region. For convenience, the term integration theorem is used whether the derivative is obtained using a del operator or a partial time derivative operator. Thus, integrals of microscopic derivatives are converted to megascopic boundary fluxes. Integration theorems comprise the set denoted:

$$\langle \text{letter} \rangle [i,\ (0, 0)\ , k] \quad \text{where} \quad 1 \le k \le 3 \quad \text{for} \quad i\ =\ 1, 2, \text{ or } 3$$

and $\langle \text{letter} \rangle$ is either G, D, C, or T indicating that the expression being transformed is a gradient, divergence, curl, or partial time derivative, respectively. No macroscopic dimensionality is considered so second and third indices in the naming convention are always zero for the integration theorems. The microscopic dimensionality of the derivative operator acting on the function is given by i. The number of megascopic dimensions that result following the integral transformation is given by k. The index k also specifies the dimensionality of the integration. Consequently, the quantity $3 - k$ is equivalent to the microscopic dimensions remaining after integration. For $i = k$ the transformation from microscopic coordinates to megascopic coordinates encompasses all of the coordinate directions of differentiation. When $k < i$, microscopic variation in $i - k$ directions of differentiation are preserved. When $k > i$, integration is performed to the megascale in more coordinate directions than those used in the differential operator.

The examples that follow are presented here to provide insight into the usage of the naming convention for the integration theorems. These examples are not intended as proofs of theorems or as the statement of the theorems. Rather, strategies for deriving theorems are presented in Chapter 6 while integration theorems are tabulated in Chapter 7.

Example: Spatial divergence theorem for a volume, D[3, (0, 0), 3]

The most commonly encountered divergence theorem is the integration over a volume of the three-dimensional divergence operator acting on a spatial vector. In this theorem, three microscopic dimensions are converted to the megascopic scale. Thus both i and k are equal to 3. No macroscopic dimensions are involved in this transformation so $m = n = 0$. This theorem for a spatial divergence integrated over a volume is thus indicated as $D[3,\ (0, 0), 3]$ and is obtained as:

D[3, (0, 0), 3]

$$\int_V \nabla \bullet \mathbf{f} dV = \int_S \mathbf{n}^* \bullet \mathbf{f} dS \tag{5.1}$$

where:

- V is the volume of integration,
- S is the surface boundary of the volume of integration,
- $\mathbf{n}^*$ is the unit normal vector outward from V on surface S,
- $\mathbf{f}$ is a spatial vector $\mathbf{f}(\mathbf{x}, t)$, and
- ∇ is the three-dimensional divergence operator.

The right side of equation (5.1) involves no derivatives of $\mathbf{f}$ but only the value of the normal component of this vector on the boundary of the domain of interest. No information about the magnitude of the divergence of the function within the volume is retained on the right side of the equation as integration has transformed all spatial dimensions to the megascopic scale. The requirement that the divergence of the vector function be known at every point within the volume has been transformed to the requirement that the normal component be known only at the edge of the megascopic domain.

Example: Spatial divergence theorem for a curve, D[3, (0, 0), 1]

Now consider a theorem where the spatial divergence of a three-dimensional vector function is to be integrated over a curve (e.g. as in figure 4.6). No macroscale dependence is to be included either along the curve or in the directions normal to the curve. This theorem is indicated as $D[3, (0, 0), 1]$ where the first index denotes that the a divergence operator in three-dimensions is being considered, the second two indices connote that no macroscale is involved, and the fourth index dictates the fact that integration to the megascale is being performed over only one spatial coordinate. The coordinate of integration is along the curve. The difference between the first and last indices, $i - k = 2$, is the number of microscopic coordinates upon which the function may depend following integration along the curve. The spatial divergence theorem $D[3, (0, 0), 1]$ is of the form:

D[3, (0, 0), 1]

$$\int_C \nabla \bullet \mathbf{f} dC = \int_C \nabla^s \bullet \mathbf{f}^s dC + \int_C (\nabla^s \bullet \boldsymbol{\lambda})\, \boldsymbol{\lambda} \bullet \mathbf{f}^c dC - \int_C (\boldsymbol{\lambda} \bullet \nabla^c \boldsymbol{\lambda}) \bullet \mathbf{f}^s dC + (\mathbf{e} \bullet \mathbf{f}^c)\big|_{\text{ends}} \tag{5.2}$$

where:

C	is the curve of integration,
$\boldsymbol{\lambda}$	is the unit vector tangent to curve C,
$\mathbf{e}$	is the unit vector tangent to curve C at the endpoints positive in the direction out from the curve,
∇^s	is the surficial operator such that $\nabla^s = \nabla - \boldsymbol{\lambda\lambda}\bullet\nabla$,
∇^c	is the curvilineal operator such that $\nabla^c = \boldsymbol{\lambda\lambda}\bullet\nabla$,
$\mathbf{f}$	is a spatial vector $\mathbf{f}(\mathbf{x}, t)$,
$\mathbf{f}^c$	is the vector tangent to C such that $\mathbf{f}^c = \boldsymbol{\lambda\lambda}\bullet\mathbf{f}$, and
$\mathbf{f}^s$	is the vector normal to C such that $\mathbf{f}^s = \mathbf{f} - \mathbf{f}^c$.

In the right side of equation (5.2), no differentiation of $\mathbf{f}$ appears along the megascopic dimension tangent to the curve. The second term on the right side of equation (5.2) accounts for the change in orientation of the curve C with position in space (i.e. $\nabla^s\bullet\boldsymbol{\lambda}$ is related to the mean curvature of the surface normal to $\boldsymbol{\lambda}$). The third term on the right side of equation (5.2) accounts for the curvature of C. The last term in the equation arises as a transformation from the derivative along the curve to an evaluation of the values of the components of $\mathbf{f}$ tangent to the curve at the two endpoints of the curve. For the special case where the curve is a closed loop, this last term would be zero.

Example: Transport theorem for a volume, T[3, (0, 0), 3]

Transport theorems transform the time derivative of a function with dependence on microscopic spatial coordinates to a time derivative of a function which is at least partially megascopic. The transport theorem to be presented in this example is the counterpart to the divergence theorem given in equation (5.1). This theorem is the general transport theorem wherein the partial time derivative of a spatial function is integrated over a volume. This theorem is specified $T[3, (0, 0), 3]$. The fact that the first index is 3, indicates that the partial time derivative is being evaluated while holding all spatial coordinates constant. The fact that the last index is 3 indicates that integration is performed over a volume, V. The transport theorem may be stated:

T[3, (0, 0), 3]

$$\int_V \left.\frac{\partial f}{\partial t}\right|_{\mathbf{x}} dV = \frac{d}{dt}\int_V f dV - \int_S f\mathbf{w}\bullet\mathbf{n}^* dS \qquad \textbf{(5.3)}$$

where:

V	is the volume of integration,
S	is the surface boundary of the volume of integration,

$\mathbf{n}^*$ is the unit normal vector outward from V on surface S,

$\mathbf{w}$ is the velocity of the surface S, and

f is a spatial function $f(\mathbf{x}, t)$.

The subscript $\mathbf{x}$ on the partial time derivative in equation (5.3) emphasizes that the partial time derivative is evaluated holding all spatial coordinates constant. By the transformation affected in equation (5.3), the partial time derivative of a microscopic function is converted to a total time derivative of a megascopic quantity plus an additional term accounting for flux across the megascopic volume boundary, S, due to the movement of this boundary.

Example: Transport theorem for a surface, T[3, (0, 0), 2]

Now consider a transport theorem for a partial time derivative with spatial coordinates held constant, such that the first index $i = 3$, where the integration is performed over an arbitrary orientable surface, such that $k = 2$. The transport theorem for this case is thus denoted $T[3, (0, 0), 2]$. Note that because $i - k = 1$, microscopic dependence on one spatial coordinate, the coordinate normal to the surface, is retained. Figure 4.5 illustrates a typical surface of integration, S, for this system. The transport theorem under consideration is written:

T[3, (0, 0), 2]

$$\int_S \left.\frac{\partial f}{\partial t}\right|_{\mathbf{x}} dS = \frac{d}{dt}\int_S f dS - \int_S (\nabla^s \cdot \mathbf{n})\, \mathbf{w}^c \cdot \mathbf{n} f dS - \int_S \mathbf{w}^c \cdot \nabla^c f dS - \int_C \boldsymbol{\nu} \cdot \mathbf{w} f dC \quad \textbf{(5.4)}$$

where:

S is the surface of integration,

C is the curve bounding the surface of integration,

$\mathbf{n}$ is the unit vector normal to surface S,

$\boldsymbol{\nu}$ is the unit vector normal to curve C such that $\mathbf{n} \cdot \boldsymbol{\nu} = 0$,

∇^c is the curvilineal operator such that $\nabla^c = \mathbf{nn} \cdot \nabla$,

∇^s is the surficial operator such that $\nabla^s = \nabla - \nabla^c$,

$\mathbf{w}$ is the velocity of S,

$\mathbf{w}^c$ is the vector component of $\mathbf{w}$ normal to the surface such that $\mathbf{w}^c = \mathbf{nn} \cdot \mathbf{w}$, and

f is a spatial function $f(\mathbf{x}, t)$,

In the right side of equation (5.4), no differentiation of f with respect to coordinates tangent to S appears. Additionally, the partial time derivative has been moved outside the integral sign and written as d/dt indicating that the time differentiation follows the moving surface S. Note that all integrals appearing on the right side of equation (5.4) are functions only of time and the single coordinate normal to the surface (consistent with the fact that $i - k = 1$). If the surface of integration is closed, such as the surface of a sphere, the last term accounting for the integration over the curve bounding the surface will be zero.

5.2.2 Averaging Theorems

As considered here, an averaging process is over the one-, two-, or three-dimensional regions contained within an averaging volume. The averaging volume, or REV, has a characteristic length in each coordinate direction corresponding to the integration scale in that direction. An averaging volume is macroscopic, megascopic, or a mix of these two scales. Specific REV's for various types of averages have been detailed in Subsection 4.2.2. Although the averaging region is a volume, the actual integration occurs over subvolumes, surfaces, or curves contained within that volume. Thus an average value of the function integrated may be associated with each REV so that even though the actual integration might be performed along a curve (i.e. in one dimension) or over a surface (i.e. in two dimensions), the averaged function will vary with REV position. Averaging theorems are not developed for volumes which are megascopic in all three spatial dimensions as these would simply be integration theorems. The REV's are macroscopic in at least one dimension and megascopic in the remaining dimensions.

In general, averaging theorems relate the integral of a derivative to the derivative of that integral. Averaging transforms a derivative of a microscopic function to a derivative of a function which depends on macroscale. Averaging theorems considered here are identified by the notation:

$$\langle \text{letter} \rangle \, [i,\ (m, n)\,,\, k]$$

where:

$\langle \text{letter} \rangle$ is either G, D, C, or T indicating that the theorem applies to a gradient, divergence, curl, or partial time derivative operator, respectively,

$i = 1, 2, \text{ or } 3$,

$m = 0, 1, 2, \text{ or } 3$,

$n = 0, 1, 2, \text{ or } 3$,

$k = 0, 1, \text{ or } 2$,

$m + n + k = 3$, and

$m \times n = 0$.

The index i, as with integration theorems, represents the microscopic dimensionality of the derivative operator acting on the function of interest. Furthermore, the magnitude of i indicates whether the integration region within the

REV is a curve ($i = 1$), a surface ($i = 2$), or space ($i = 3$) and the minimum dimensionality of the function.

The meanings of the second and third indices, m and n, are subtle. The averaging theorems considered here are restricted such that either m or n will be zero. When $m > 0$, macroscale derivatives of averaged quantities are generated through application of the averaging theorem. On the other hand, when $n > 0$, the dependence of the resulting terms on REV position is through integration over the curve, surface, or volume of integration intersecting the boundary of the REV. The case where $m > 0$ seems to be of greater interest, but useful corollaries to the basic theorems are obtained by equating the right sides of $\langle\text{letter}\rangle[i,\ (m, 0),\ k]$ to the right sides of $\langle\text{letter}\rangle[i,\ (0, n),\ k]$ when $m = n$.

The examples that follow provide insight into the notational convention used to designate averaging theorems. These examples are not proofs of the theorems. Strategies for theorem derivation are presented in Chapter 6. Averaging theorems themselves are tabulated in Chapter 8.

Example: Averaging theorem for spatial divergence, D[3, (3, 0), 0]

The averaging theorem in three dimensions given by $D[3,\ (3, 0),\ 0]$ is concerned with integration of the spatial divergence of a vector function over a portion of space within a fully macroscopic REV. Let the portion of the REV that is to be integrated over be that part containing, for example, the α-phase of a multiphase system. When the REV size, shape, and orientation is independent of space and time, the theorem obtained is:

D[3, (3, 0), 0]

$$\int_{V_\alpha} \nabla\cdot\mathbf{f}\,dV = \nabla\cdot\int_{V_\alpha}\mathbf{f}\,dV + \int_{S_{\alpha\beta}}\mathbf{n}\cdot\mathbf{f}\,dS \qquad (5.5)$$

where:

- V_α is the portion of the REV occupied by the α-phase,
- $S_{\alpha\beta}$ is the boundary surface within the REV between the α-phase and all other phases,
- $\mathbf{n}$ is the unit normal vector on the $S_{\alpha\beta}$ surface of integration pointing outward from the α-phase,
- ∇ in the integrand is the microscopic spatial operator, $\nabla = \nabla_\xi$,
- ∇ outside the integral is the macroscopic spatial operator, $\nabla = \nabla_\mathbf{x}$, and
- $\mathbf{f}$ is a spatial vector $\mathbf{f}(\mathbf{x}, t)$.

Of importance is the fact that the divergence operator on the left side of equation (5.5) is with respect to microscopic coordinates while the divergence operator on the right side is with respect to macroscopic coordinates. If the α-phase is the only phase present, there will be no boundary surfaces within the REV; and the last integral in equation (5.5) will not arise.

Example: Averaging theorem for spatial divergence, D[3, (0, 3), 0]

A somewhat different averaging theorem from that considered in the last example is $D[3, (0, 3), 0]$. Note that this theorem has the same first and last indices as the previous case indicating that both transform a spatial divergence operator at the microscale to a form that is macroscopic. However, in the form of interest in the current example, the macroscale relations are treated using surface integrals over the averaging volume. Because the second index is zero, no macroscopic spatial derivatives of the function being averaged should appear in the final form. The form obtained for this theorem is:

D[3, (0, 3), 0]

$$\int_{V_\alpha} \nabla \bullet \mathbf{f} dV = \int_{S_\alpha} \mathbf{n}^* \bullet \mathbf{f} dS + \int_{S_{\alpha\beta}} \mathbf{n} \bullet \mathbf{f} dS \tag{5.6}$$

where:

- V_α is the portion of the REV occupied by the α-phase,
- S_α is the portion of the external boundary of the REV that intersects the α-phase,
- $S_{\alpha\beta}$ is the boundary surface within the REV between the α-phase and all other phases,
- $\mathbf{n}^*$ is the unit vector normal to the external boundary of the REV pointing outward from the REV,
- $\mathbf{n}$ is the unit normal vector on the $S_{\alpha\beta}$ surface of integration pointing outward from the α-phase,
- ∇ is the microscopic spatial operator, $\nabla = \nabla_\xi$, and
- $\mathbf{f}$ is a spatial vector $\mathbf{f}(\mathbf{x}, t)$.

The averaging theorem of equation (5.6) converts the volume integral of variations at the microscale to integrals over surfaces on the boundary of the REV and within the REV. The values of these surface integrals are macroscopic quantities associated in space with the centroid of each REV. If the α-phase is the only phase present, the first integral on the right side of equation (5.6) will be over the entire boundary surface of the REV, and the last integral will not arise.

Example: Corollary to D[3, (3, 0), 0] and D[3, (0, 3), 0]

The left sides of the averaging theorems presented as equations (5.5) and (5.6) are identical. Therefore, the right sides must be equal to each other. This observation requires that the following identity hold:

$$\nabla \bullet \int_{V_\alpha} \mathbf{f} dV = \int_{S_\alpha} \mathbf{n}^* \bullet \mathbf{f} dS \tag{5.7}$$

Corollaries of this type offer some insight to the mathematical relations derived as averaging theorems.

Example: Averaging theorem for surface divergence, D[2, (1, 0), 2]

The averaging theorem $D[2, (1, 0), 2]$ is presented as a demonstration of a theorem that combines macroscopic and megascopic averaging. The first index, 2, coupled with the designation, D, indicates that the theorem applies to the surface divergence of a vector. Integration will be performed over simple surfaces contained within the averaging volume. The last index, 2, indicates that the averaging volume is megascopic in two dimensions. The fact that the second index in parentheses is zero indicates that the macroscopic dependence will be treated as a derivative with respect to the macroscopic coordinate rather than as an integral over the macroscopic surface of the REV. A typical averaging volume for this case is presented in figure 4.10 (or figure 4.11). Note that the vector **N** is a unit vector defined at the macroscale to be orthogonal to the face of the averaging slab and tangent to the macroscopic coordinate. When the averaging is completed, the macroscopic function retains its spatial dependence on the location of the averaging volume. A macroscopic one-dimensional divergence operator along the curve accounts for the spatial variation along the curve. For this case, the averaging theorem is:

D[2, (1, 0), 2]

$$\int_{S_{\alpha\beta}} \nabla^s \bullet \mathbf{f} dS = \nabla^C \bullet \int_{S_{\alpha\beta}} \mathbf{f}^s dS + \int_{S_{\alpha\beta}} (\nabla^s \bullet \mathbf{n})\, \mathbf{n} \bullet \mathbf{f}^c dS + \int_{C_{\alpha\beta\epsilon}} \boldsymbol{\nu} \bullet \mathbf{f}^s dC + \int_{C_{\alpha\beta_{\text{edges}}}} \boldsymbol{\nu} \bullet \mathbf{f}^s dC \tag{5.8}$$

where:

- $S_{\alpha\beta}$ is the surface within the slab REV between the α-phase and all other phases,
- $C_{\alpha\beta\epsilon}$ is the boundary curve of $S_{\alpha\beta}$ within the REV that also is the location where the α- and β-phases meet a third phase,

$C_{\alpha\beta_{\text{edges}}}$ is the curve at the edge of the REV formed by the intersection of $S_{\alpha\beta}$ with the edge of the slab,

n is the unit vector normal to the $S_{\alpha\beta}$ surface,

$\boldsymbol{\nu}$ is the unit vector tangent to the $S_{\alpha\beta}$ surface and normal to $C_{\alpha\beta\epsilon}$ and $C_{\alpha\beta_{\text{edge}}}$ such that $\mathbf{n}\bullet\boldsymbol{\nu} = 0$,

N is the unit vector normal to the face of the slab REV,

∇ is the microscopic spatial operator, $\nabla = \nabla_{\xi}$,

∇^s is the microscopic surficial operator, $\nabla^s = \nabla_{\xi} - \mathbf{nn}\bullet\nabla_{\xi}$,

∇^C is the one-dimensional macroscopic spatial operator in the direction normal to the face of the slab REV, $\nabla^C = \mathbf{NN}\bullet\nabla_{\mathbf{x}}$,

f is a spatial vector $\mathbf{f}(\mathbf{u}, t)$ defined on $S_{\alpha\beta}$,

$\mathbf{f}^c$ is the vector component of **f** normal to $S_{\alpha\beta}$, $\mathbf{f}^c = \mathbf{nn}\bullet\mathbf{f}$, and

$\mathbf{f}^s$ is the vector component of **f** tangent to $S_{\alpha\beta}$, $\mathbf{f}^s = \mathbf{f} - \mathbf{f}^c$.

When the REV contains only two distinct regions, such as an α-phase and a β-phase, no curve $C_{\alpha\beta\epsilon}$ will exist and the integral over this curve in equation (5.8) will not arise. Furthermore, if the vector **f** has no component normal to the surface such that $\mathbf{n}\bullet\mathbf{f} = 0$, **f** in the equation may be replace by $\mathbf{f}^s$; and $\mathbf{f}^c$ will equal zero.

5.3 CONCLUSION

The purpose of this chapter has been to provide the rationale behind the method of identification of the averaging and integration theorems. This is important because of the large number of theorems (128 in all) that are going to be discussed. If the theorems are to be useful in a particular application, one must be able to find the appropriate theorem for that application based on the desired attributes of the result to be obtained. The scheme employed here is actually rather straightforward if one notes that the first index refers to the dimensionality of the microscopic differential operator, the sum of the second two indices is the number of macroscopic coordinates in the transformed expression, and the last index provides the number of megascopic dimensions included. A summary of the theorem classification scheme is presented in Table 5.1. The examples presented here are not intended to provide insight into the method of theorem derivation based on generalized functions, but only insight into the features of the classification scheme. The next chapter will begin the presentation of the mathematical tools which are employed in theorem derivation.

Table 5.1: Summary of theorem classifications and notation

THEOREM CLASS	NOTATION	VALUES OF INDICES	COMMENTS
Integration	⟨letter⟩[i, (0, 0), k]	$i = 1, 2$, or 3 $k = 1, 2$, or 3	No macroscopic dimensions. i is the dimensionality of the differential operator. k is the dimensionality of the integration.
Averaging	⟨letter⟩[i, (m, n), k]	$i = 1, 2$, or 3 $m = 0, 1, 2$, or 3 $n = 0, 1, 2$, or 3 $k = 0, 1$, or 2 $m + n + k = 3$ $m \times n = 0$	No microscopic dimensions after transformation. i is the dimensionality of the microscopic differential operator. Integration over i-dimensional space within an REV. k is the number of megascopic dimensions. $m + n$ megascopic dimensions after integration. m-dimensional macroscopic differentiation. n-dimensional integration over boundary of an REV.

CHAPTER SIX

APPLICATIONS OF GENERALIZED FUNCTIONS

6.0 INTRODUCTION

The concept of using generalized functions in the derivation of integral theorems is straightforward. By application of the rules developed in Chapter 4, line integrals, surface integrals and volume integrals can be transformed to integrals over all space with the generalized functions and/or their derivatives appearing in the integrand. Subsequent manipulations involving generalized functions, required to arrive at concise and acceptable formulations of the integral theorems, are extensive in some cases. However, when compared to alternative derivations, they are reasonably simple and easy to follow. Repeated application of the chain rule for differentiation is required to convert integrals of space or time derivatives to derivatives of integrals and to eliminate the generalized functions and their derivatives from the resulting expressions. Fortunately, a number of rules that are generally valid can be applied to reduce the amount of work required in deriving these theorems.

In this chapter, these rules or identities will be given or derived. It should be understood that the identities will be applied in the theorem derivation only under integration. Detailed derivations of the Gauss divergence theorem $D[3, (0,0), 3]$, the averaging theorems G, D, C, and $T[3, (1,0), 2]$, and the transport theorem $T[2, (0,0), 1]$ will be provided to illustrate the application of these rules. The $[3, (1,0), 2]$ family of averaging theorems describes the integration of three-dimensional microscopic space and time derivatives to a system that is macroscopically one-dimensional and megascopically two-dimensional. The transport theorem $T[2, (0,0), 1]$ provides a relation between the integral over a surface of a partial time derivative of a function and the time derivative of the surface integral of the function.

6.1 SOME USEFUL IDENTITIES INVOLVING GENERALIZED FUNCTIONS

Although the use of generalized functions greatly simplifies the derivation of the various integral theorems, a significant degree of manipulation of generalized functions is required to obtain the final forms of the theorems. A large number of individual terms involving integrals of expressions containing spatial and/or temporal derivatives of generalized functions will be encountered during the derivations. Although many of these expressions are not readily converted to integrals over regions of reduced dimensionality, generally valid identities for others can be derived that are helpful in reducing the amount of work required in theorem derivation. These identities involve expressions containing the derivatives of volumetric, surficial, and curvilineal generalized functions and expressions in the unit vectors $\boldsymbol{\lambda}$, $\mathbf{n}$, and $\boldsymbol{\nu}$ defining a general right-handed orthogonal coordinate system. In the following subsections, a number of useful identities will be presented.

6.1.1 Orthogonality Relations

For expressions that contain combinations of the time or space derivative of the unit vector $\mathbf{n}$, $\boldsymbol{\nu}$, or $\boldsymbol{\lambda}$ and the space derivative of the associated generalized function γ, γ^s, or γ^c it can be shown that:

Identity 6.1a

$$\frac{\partial \mathbf{n}}{\partial t} \bullet \nabla \gamma = 0 \tag{6.1a}$$

Identity 6.1b

$$\nabla \mathbf{n} \bullet \nabla \gamma = 0 \tag{6.1b}$$

where:

- γ is the generalized function used to identify a volume in space,
- $\mathbf{n}$ is the unit vector normal to the surface of the volume, and
- ∇ is the spatial gradient operator.

Identity 6.1c

$$\frac{\partial \boldsymbol{\nu}}{\partial t} \bullet \nabla^s \gamma^s = \frac{\partial \boldsymbol{\nu}}{\partial t} \bullet \nabla \gamma^s = 0 \tag{6.1c}$$

Identity 6.1d

$$\nabla \boldsymbol{\nu} \bullet \nabla^s \gamma^s = \nabla \boldsymbol{\nu} \bullet \nabla \gamma^s = 0 \tag{6.1d}$$

where:

- γ^s is the generalized function used to identify a simple surface that is portion of a much larger surface,
- $\mathbf{n}$ is a unit vector normal to the simple surface,
- $\boldsymbol{\nu}$ is a unit vector tangent to the simple surface such that $\boldsymbol{\nu}\bullet\mathbf{n} = 0$ and also normal to the curve that describes the boundary of this surface,
- ∇ is the spatial gradient operator, and
- ∇^s is the surface gradient operator, $\nabla^s = \nabla - \mathbf{nn}\bullet\nabla$.

Identity 6.1e

$$\frac{\partial\boldsymbol{\lambda}}{\partial t}\bullet\nabla^c\gamma^c = \frac{\partial\boldsymbol{\lambda}}{\partial t}\bullet\nabla\gamma^c = 0 \qquad \textbf{(6.1e)}$$

Identity 6.1f

$$\nabla\boldsymbol{\lambda}\bullet\nabla^c\gamma^c = \nabla\boldsymbol{\lambda}\bullet\nabla\gamma^c = 0 \qquad \textbf{(6.1f)}$$

where:

- γ^c is the generalized function used to identify a simple curve that is a segment of a longer curve,
- $\boldsymbol{\lambda}$ is the unit vector tangent to the simple curve,
- ∇ is the spatial gradient operator, and
- ∇^c is the curvilineal gradient operator $\nabla^c = \boldsymbol{\lambda\lambda}\bullet\nabla$.

Equations (6.1a) through (6.1f) are proven easily making use of equations (2.12) and (2.13) and the facts that $\nabla\gamma$, $\nabla^s\gamma^s$, and $\nabla^c\gamma^c$ are collinear with $\mathbf{n}$, $\boldsymbol{\nu}$, and $\boldsymbol{\lambda}$, respectively. As an example, the proof of equation (6.1d) will be presented here.

The more difficult part of the proof of equation (6.1d) concerns the term $\nabla\boldsymbol{\nu}\bullet\nabla\gamma^s$. Application of the chain rule yields:

$$\nabla\boldsymbol{\nu}\bullet\nabla\gamma^s = \nabla(\boldsymbol{\nu}\bullet\nabla\gamma^s) - \boldsymbol{\nu}\bullet\nabla\nabla\gamma^s \qquad \textbf{(6.2)}$$

Because equation (2.26) indicates that the surficial operator is given by:

$$\nabla^s = \nabla - \mathbf{nn}\bullet\nabla \qquad \textbf{(6.3)}$$

the first term on the right side of equation (6.2) can be written as:

$$\nabla(\boldsymbol{\nu}\bullet\nabla\gamma^s) = \nabla(\boldsymbol{\nu}\bullet\nabla^s\gamma^s + \boldsymbol{\nu}\bullet\mathbf{nn}\bullet\nabla\gamma^s) \qquad \textbf{(6.4a)}$$

or, since $\boldsymbol{\nu}$ and $\mathbf{n}$ are orthogonal:

$$\nabla(\boldsymbol{\nu}\cdot\nabla\gamma^s) = \nabla(\boldsymbol{\nu}\cdot\nabla^s\gamma^s) \tag{6.4b}$$

Substitution of equation (6.4b) into equation (6.2), differentiation of the resulting expression, and rearrangement gives:

$$\nabla\boldsymbol{\nu}\cdot\nabla\gamma^s = \nabla\boldsymbol{\nu}\cdot\nabla^s\gamma^s + \boldsymbol{\nu}\cdot[(\nabla^s - \nabla)\nabla\gamma^s] \tag{6.5}$$

Equation (6.3) indicates that $\nabla^s - \nabla = -\mathbf{nn}\cdot\nabla$ so that:

$$\nabla\boldsymbol{\nu}\cdot\nabla\gamma^s = \nabla\boldsymbol{\nu}\cdot\nabla^s\gamma^s - \boldsymbol{\nu}\cdot\mathbf{nn}\cdot\nabla\nabla\gamma^s \tag{6.6}$$

The first term on the right side of this equation is zero because, by equation (3.30), $\nabla^s\gamma^s$ is a vector in the $\boldsymbol{\nu}$ direction; and, by equation (2.12), $\nabla\boldsymbol{\nu}\cdot\boldsymbol{\nu} = 0$. The second term is zero because $\boldsymbol{\nu}$ and $\mathbf{n}$ are orthogonal. Thus equation (6.1d) is demonstrated to hold.

6.1.2 Gradients of Generalized Functions

Consider the transformation of an integral along a closed curve C that is the boundary of a surface S to an integral over a surface S_∞ containing the surface S. This transformation makes use of the surface gradient of the generalized function γ^s in the surface S_∞, giving rise to the expression, discussed as equation (3.32):

$$\int_C f dC = -\int_{S_\infty} f\boldsymbol{\nu}\cdot\nabla^s\gamma^s dS \tag{6.7}$$

where:

- $\boldsymbol{\nu}$ is a unit normal vector to the closed curve C in the surface S_∞,
- ∇^s is the surface gradient in the surface S_∞, and
- γ^s is a generalized function for the surface S_∞ that is equal to 1 in S but is 0 elsewhere in S_∞ such that it undergoes a change in value on the curve C.

Subsequent manipulations of expressions such as that on the right side of equation (6.7) are sometimes more easily carried to a useful conclusion if the surficial operator, ∇^s is related to the spatial operator ∇.

Similarly, repeated application of the chain rule during theorem derivation often leads to expressions containing curvilineal or surficial gradients of the generalized function for a curve, $\nabla^c\gamma^c$ and $\nabla^s\gamma^c$, respectively. Manipulation of these terms is more easily carried out if the curvilineal or surficial gradients of a generalized function are related to the spatial gradient. Such relations can be obtained from the orthogonality properties given by equations (6.1a) through (6.1f).

Consider the identity given in equation (6.1c):

$$\frac{\partial \boldsymbol{\nu}}{\partial t} \bullet \nabla^s \gamma^s = \frac{\partial \boldsymbol{\nu}}{\partial t} \bullet \nabla \gamma^s = 0 \qquad \textbf{(6.8)}$$

This equation may be rearranged to the form:

$$\frac{\partial \boldsymbol{\nu}}{\partial t} \bullet (\nabla - \nabla^s)\, \gamma^s = 0 \qquad \textbf{(6.9)}$$

Now substitution of equation (6.3) into this expression yields:

$$\frac{\partial \boldsymbol{\nu}}{\partial t} \bullet \mathbf{n}\mathbf{n} \bullet \nabla \gamma^s = 0 \qquad \textbf{(6.10)}$$

Equation (6.10) is valid for any simple surface, S, and closed curve, C, in the surface associated with the generalized function γ^s. However, since $(\partial \boldsymbol{\nu}/\partial t) \bullet \mathbf{n}$ is in general non-zero, it follows that:

Identity 6.2a

$$\mathbf{n} \bullet \nabla \gamma^s = 0 \qquad \textbf{(6.11a)}$$

where:

- γ^s is a generalized function for a surface, and
- $\mathbf{n}$ is the normal to the surface under consideration.

In a similar manner, the following identity can be proven:

Identity 6.2b

$$\boldsymbol{\nu} \bullet \nabla \gamma^c = \mathbf{n} \bullet \nabla \gamma^c = 0 \qquad \textbf{(6.11b)}$$

where:

- γ^c is a generalized function for a curve, and
- $\mathbf{n}$ is a unit vector normal to the curve, and
- $\boldsymbol{\nu}$ is another unit vector normal to the curve but also normal to $\mathbf{n}$ such that $\mathbf{n} \bullet \boldsymbol{\nu} = 0$.

Equations (6.11a) and (6.11b), respectively, can be used to derive the following useful identities:

Identity 6.3a

$$\nabla \gamma^s = \nabla^s \gamma^s \qquad \textbf{(6.12a)}$$

Identity 6.3b

$$\nabla \gamma^{c} = \nabla^{s} \gamma^{c} = \nabla^{c} \gamma^{c} \tag{6.12b}$$

These last two identities are extremely useful in that they indicate a direct change in the type of gradient operator acting on the generalized function for a surface or for a curve is valid.

The dyad obtained as the product of the vector $\mathbf{n}$ normal to a surface S and the gradient of the spatial generalized function γ which changes value on the surface can, with the aid of equation (3.37), be written as:

$$\mathbf{n} \nabla \gamma = -\mathbf{n n} \delta [g(\mathbf{x}; \mathbf{x}^{s})] \tag{6.13}$$

where:

$g(\mathbf{x}; \mathbf{x}^{s})$ is zero on the surface, and

$\delta [g(\mathbf{x}; \mathbf{x}^{s})]$ is the Dirac delta function for the surface S.

The right side of equation (6.13) is obviously a symmetric tensor, so the left side must also be symmetric such that the following identity holds:

Identity 6.4a

$$\mathbf{n} \nabla \gamma = (\nabla \gamma) \mathbf{n} \tag{6.14a}$$

By similar reasoning, based on equations (3.30) and (3.24), and by invoking equations (6.12a) and (6.12b), the following identities are also obtained:

Identity 6.4b

$$\boldsymbol{\nu} \nabla^{s} \gamma^{s} = (\nabla^{s} \gamma^{s}) \boldsymbol{\nu} \tag{6.14b}$$

Identity 6.4c

$$\boldsymbol{\nu} \nabla \gamma^{s} = (\nabla \gamma^{s}) \boldsymbol{\nu} \tag{6.14c}$$

Identity 6.4d

$$\boldsymbol{\lambda} \nabla^{c} \gamma^{c} = (\nabla^{c} \gamma^{c}) \boldsymbol{\lambda} \tag{6.14d}$$

Identity 6.4e

$$\boldsymbol{\lambda} \nabla^{s} \gamma^{c} = (\nabla^{s} \gamma^{c}) \boldsymbol{\lambda} \tag{6.14e}$$

Identity 6.4f

$$\boldsymbol{\lambda} \nabla \gamma^{c} = (\nabla \gamma^{c}) \boldsymbol{\lambda} \tag{6.14f}$$

In these relations, the notation and unit vectors are the same as indicated following equations (6.1b), (6.1d), and (6.1f).

6.1.3 Time Derivatives of Generalized Functions

The partial time derivative for a spatial generalized function γ is evaluated holding the spatial coordinates constant. On the other hand, the partial time derivative of a surficial generalized function γ^s on a surface S_∞ is obtained keeping the surface coordinates constant. Similarly, the partial time derivative of a generalized function for a curve, γ^c, is computed keeping the coordinate along the curve constant. However, in the manipulations required in deriving integral theorems, expressions for the partial time derivatives of γ^s and γ^c with spatial coordinates $\mathbf{x}$ held constant are convenient. These may be obtained as illustrated here.

First, equation (3.35) is:

$$\left.\frac{\partial \gamma^s}{\partial t}\right|_{\mathbf{u}} = -\mathbf{w}^s \cdot \nabla^s \gamma^s \tag{6.15}$$

where:

$\mathbf{w}^s$ is the velocity of the surface in the directions tangent to the surface, and

∇^s is the surface gradient operator, $\nabla - \mathbf{nn} \cdot \nabla$.

By virtue of equation (6.12a), the surface gradient operator on the right side of equation (6.15) may be replaced by the spatial gradient operator to obtain:

$$\left.\frac{\partial \gamma^s}{\partial t}\right|_{\mathbf{u}} = -\mathbf{w}^s \cdot \nabla \gamma^s \tag{6.16}$$

Since, by definition, $\mathbf{w}^s = \mathbf{w} - \mathbf{nn} \cdot \mathbf{w}$, and because $\mathbf{n} \cdot \nabla \gamma^s = 0$ by equation (6.11a), equation (6.16) may also be written:

$$\left.\frac{\partial \gamma^s}{\partial t}\right|_{\mathbf{u}} = -\mathbf{w} \cdot \nabla \gamma^s \tag{6.17}$$

Equation (2.68) provides the relation between the time derivative of a function evaluated keeping the space coordinates constant and the time derivative keeping the surface coordinates constant. If this function is taken to be γ^s:

$$\left.\frac{\partial \gamma^s}{\partial t}\right|_{\mathbf{x}} = \left.\frac{\partial \gamma^s}{\partial t}\right|_{\mathbf{u}} - \mathbf{w} \cdot \mathbf{nn} \cdot \nabla \gamma^s \tag{6.18}$$

However, by equation (6.11a), the last term in this expression will be zero. Thus combination of equations (6.17) and (6.18) results in the following identity:

Identity 6.5a

$$\left.\frac{\partial \gamma^s}{\partial t}\right|_{\mathbf{x}} = \left.\frac{\partial \gamma^s}{\partial t}\right|_{\mathbf{u}} = -\mathbf{w} \cdot \nabla \gamma^s \qquad \textbf{(6.19a)}$$

In a similar manner, the following identity for the time derivatives of the curvilineal generalized function γ^c can be derived:

Identity 6.5b

$$\left.\frac{\partial \gamma^c}{\partial t}\right|_{\mathbf{x}} = \left.\frac{\partial \gamma^c}{\partial t}\right|_{\mathbf{u}} = \left.\frac{\partial \gamma^c}{\partial t}\right|_{\mathbf{l}} = -\mathbf{w} \cdot \nabla \gamma^c \qquad \textbf{(6.19b)}$$

These last two equations are extremely useful in that they allow for a direct change in the type of temporal differential operator acting on the generalized function for a surface or for a curve or for a change from the temporal operator to a gradient operator.

6.1.4 Integrands Containing the Del Operator

The integral of a spatial derivative of a vector or scalar quantity over a finite volume V may be transformed to an integral over an infinite volume V_∞ by using the appropriate spatial generalized function γ. Similarly, the integral over a finite surface of the surface del operator acting on a vector or scalar may be transformed to an integral over an infinite surface by using the generalized function γ^s. Finally, the use of the generalized function γ^c allows the integral along a finite curve of the curvilineal del operator acting on a vector or scalar to be transformed to the integral over an infinite curve. In the derivation of integral theorems, the chain rule is commonly applied such that, for example, an integrand of the form $\gamma \nabla f$ is converted to $\nabla (\gamma f) - f \nabla \gamma$. Similar expressions are also obtained when applying the chain rule to expressions involving surface and curvilineal gradients and generalized functions.

In the derivation of some averaging theorems, differentiation of an integrand will be with respect to macroscopic coordinates while integration will be over microscopic coordinates. In these cases, when the limits of integration are independent of the macroscopic coordinates, the order of differentiation and integration may be directly interchanged. Thus the integral of a derivative becomes the derivative of an integral. However, for the integration theorems and other averaging theorems, integration and differentiation are with respect to the same coordinate system. In these cases, the differential operator cannot be moved across the integral sign, but the following identities can be shown to hold:

Identity 6.6a

$$\int_{V_\infty} \nabla_\xi (\gamma f)\, dV_\xi = 0 \tag{6.20a}$$

Identity 6.6b

$$\int_{V_\infty} \nabla_\xi \bullet (\gamma \mathbf{f})\, dV_\xi = 0 \tag{6.20b}$$

Identity 6.6c

$$\int_{V_\infty} \nabla_\xi \times (\gamma \mathbf{f})\, dV_\xi = 0 \tag{6.20c}$$

Identity 6.6d

$$\int_{S_\infty} \nabla_{\mathbf{u}}^s (\gamma^s f^s)\, dS_{\mathbf{u}} = 0 \tag{6.20d}$$

Identity 6.6e

$$\int_{S_\infty} \nabla_{\mathbf{u}}^s \bullet (\gamma^s \mathbf{f}^s)\, dS_{\mathbf{u}} = 0 \tag{6.20e}$$

Identity 6.6f

$$\int_{S_\infty} \nabla_{\mathbf{u}}^s \times (\gamma^s \mathbf{f}^s)\, dS_{\mathbf{u}} = 0 \tag{6.20f}$$

Identity 6.6g

$$\int_{C_\infty} \nabla_{\mathbf{l}}^c (\gamma^c f^c)\, dC_{\mathbf{l}} = 0 \tag{6.20g}$$

Identity 6.6h

$$\int_{C_\infty} \nabla_{\mathbf{l}}^c \bullet (\gamma^c \mathbf{f}^c)\, dC_{\mathbf{l}} = 0 \tag{6.20h}$$

Identity 6.6i

$$\int_{C_\infty} \nabla_{\mathbf{l}}^c \times (\gamma^c \mathbf{f}^c)\, dC_{\mathbf{l}} = 0 \tag{6.20i}$$

where the subscripts $\boldsymbol{\xi}$, $\mathbf{u}$, and $\mathbf{l}$ are explicitly used here to denote the fact that differentiation and integration are carried out in the same coordinate system. In subsequent chapters, when the chance for confusion is reduced, these subscripts will be dropped. However capitalization will be used such that ∇^S and ∇^C will indicate surface and curvilineal del operators with respect to macroscopic coordinates, respectively, while ∇^s and ∇^c will indicate the corresponding del operators with respect to microscopic coordinates. Proofs of equations (6.20b) and (6.20e) follow.

In considering equation (6.20b), use cartesian coordinates and let $\mathbf{f}$ be the vector (f_ξ, f_η, f_ζ). Note that the generalized function γ is defined to be equal to one in some part or parts of space and zero elsewhere. The "boundaries" of V_∞ are at infinity and γ is zero there. Now expand the integrand on the left side of equation (6.20b) such that:

$$\int_{V_\infty} \nabla_\xi \bullet (\gamma \mathbf{f})\, dV_\xi = \int_{-\infty}^{\infty}\int_{-\infty}^{\infty}\int_{-\infty}^{\infty} \left[\frac{\partial(\gamma f_\xi)}{\partial \xi} + \frac{\partial(\gamma f_\eta)}{\partial \eta} + \frac{\partial(\gamma f_\zeta)}{\partial \zeta}\right] d\xi\, d\eta\, d\zeta \tag{6.21}$$

Each of the three terms on the right side of equation (6.21) may be integrated directly to obtain:

$$\int_{V_\infty} \nabla_\xi \bullet (\gamma \mathbf{f})\, dV_\xi = \int_{-\infty}^{\infty}\int_{-\infty}^{\infty} (\gamma f_\xi)\Big|_{\xi=-\infty}^{\xi=\infty} d\eta\, d\zeta + \int_{-\infty}^{\infty}\int_{-\infty}^{\infty} (\gamma f_\eta)\Big|_{\eta=-\infty}^{\eta=\infty} d\xi\, d\zeta$$

$$+ \int_{-\infty}^{\infty}\int_{-\infty}^{\infty} (\gamma f_\zeta)\Big|_{\zeta=-\infty}^{\zeta=\infty} d\xi\, d\eta \tag{6.22}$$

Because γ is zero when any coordinate equals $\pm\infty$, all terms on the right side of equation (6.22) are zero. Thus equation (6.20b) is obtained.

The proof of equation (6.20e) makes use of the power of generalized functions in transforming a surface integral to a volume integral. Consider S_∞ to be a large closed surface for which $\gamma^s = 1$ on some portion, and $\gamma^s = 0$ on the rest. The unit normal vector pointing outward from S_∞ is $\mathbf{n}$. Now consider a generalized function in space, γ, defined such that $\gamma = 1$ within S_∞ and $\gamma = 0$ in the rest of space. Equation (3.39) may be applied to change the left side of equation (6.20e) to an integral over V_∞ such that:

$$\int_{S_\infty} \nabla^s_{\mathbf{u}} \bullet (\gamma^s \mathbf{f}^s)\, dS_{\mathbf{u}} = -\int_{V_\infty} \nabla^s_{\mathbf{u}} \bullet (\gamma^s \mathbf{f}^s)\, \mathbf{n} \bullet \nabla_\xi \gamma dV_\xi \tag{6.23}$$

The surface divergence of a surface vector may be converted to a spatial divergence of a space vector by applying equation (2.41) to yield:

$$\int_{S_\infty} \nabla^s_{\mathbf{u}} \bullet (\gamma^s \mathbf{f}^s)\, dS_{\mathbf{u}} = -\int_{V_\infty} \nabla_\xi \bullet (\gamma^s \mathbf{f})\, \mathbf{n} \bullet \nabla_\xi \gamma dV_\xi + \int_{V_\infty} \nabla_\xi \bullet (\gamma^s \mathbf{nf}) \bullet \mathbf{nn} \bullet \nabla_\xi \gamma dV_\xi \tag{6.24}$$

The chain rule is next applied to both integrands on the right side of this expression to provide:

$$\int_{S_\infty} \nabla^s_{\mathbf{u}} \bullet (\gamma^s \mathbf{f}^s)\, dS_{\mathbf{u}} = -\int_{V_\infty} \nabla_\xi \bullet (\gamma^s \mathbf{fn} \bullet \nabla_\xi \gamma)\, dV_\xi + \int_{V_\infty} \gamma^s \mathbf{f} \bullet \nabla_\xi (\mathbf{n} \bullet \nabla_\xi \gamma)\, dV_\xi$$
$$+ \int_{V_\infty} \nabla_\xi \bullet (\gamma^s \mathbf{nf} \bullet \mathbf{nn} \bullet \nabla_\xi \gamma)\, dV_\xi - \int_{V_\infty} \gamma^s \mathbf{n} \bullet \nabla_\xi (\mathbf{nn} \bullet \nabla_\xi \gamma) \bullet \mathbf{f} dV_\xi \tag{6.25}$$

From equation (6.20b), the first and third integrals on the right side of this equation are zero. Based on equation (6.14a), $\mathbf{nn} \bullet \nabla_\xi \gamma = \nabla_\xi \gamma$. Substitution of this expression into the fourth integral and expanding out the second integral in equation (6.25) gives:

$$\int_{S_\infty} \nabla^s_{\mathbf{u}} \bullet (\gamma^s \mathbf{f}^s)\, dS_{\mathbf{u}} = \int_{V_\infty} \gamma^s \mathbf{f} \bullet \nabla_\xi \mathbf{n} \bullet \nabla_\xi \gamma dV_\xi + \int_{V_\infty} \gamma^s \mathbf{f} \bullet (\nabla_\xi \nabla_\xi \gamma) \bullet \mathbf{n} dV_\xi$$
$$- \int_{V_\infty} \gamma^s \mathbf{n} \bullet (\nabla_\xi \nabla_\xi \gamma) \bullet \mathbf{f} dV_\xi \tag{6.26}$$

By equation (6.1b), $\nabla_\xi \mathbf{n} \bullet \nabla_\xi \gamma = 0$ so that the first integral on the right side of equation (6.26) is zero. The integrands of the second and third integrals are equal and therefore these two integrals cancel leaving equation (6.20e) as the result.

The proof of equation (6.20h) is similar to the proof of equation (6.20e) in that the integral over a curve is converted to an integral over a surface using equation (4.10). Subsequent manipulations that parallel those above lead to the required identity. Alternatively, one could convert the integral over a curve to an integral over a volume using equation (4.11) and perform the manipulations on the expression obtained. Similar procedures to prove the identities involving the gradient and curl operators may be employed.

6.1.5 Dyadic Del of Generalized Functions

In deriving integral theorems, the transformation from integration along a curve C or over a surface S to integration over an infinite volume V_∞ makes use of the generalized functions γ, γ^S, and γ^C for a volume, surface and curve, respectively. After the chain rule is applied and some of the identities already presented are invoked, tensor terms involving the dyad $\nabla\nabla$ operating on generalized functions may arise. The following identities can be derived for these tensor terms, where the del operator may use either microscopic or macroscopic coordinates:

Identity 6.7a

$$(\nabla\nabla\gamma) \bullet \boldsymbol{\nu} = -\nabla\boldsymbol{\nu} \bullet \nabla\gamma \qquad \textbf{(6.27a)}$$

Identity 6.7b

$$(\nabla\nabla\gamma) \bullet \boldsymbol{\lambda} = -\nabla\boldsymbol{\lambda} \bullet \nabla\gamma \qquad \textbf{(6.27b)}$$

Identity 6.7c

$$(\nabla\nabla\gamma^S) \bullet \mathbf{n} = -\nabla\mathbf{n} \bullet \nabla\gamma^S \qquad \textbf{(6.27c)}$$

Identity 6.7d

$$(\nabla\nabla\gamma^S) \bullet \boldsymbol{\lambda} = -\nabla\boldsymbol{\lambda} \bullet \nabla\gamma^S \qquad \textbf{(6.27d)}$$

Identity 6.7e

$$(\nabla\nabla\gamma^C) \bullet \mathbf{n} = -\nabla\mathbf{n} \bullet \nabla\gamma^C \qquad \textbf{(6.27e)}$$

Identity 6.7f

$$(\nabla\nabla\gamma^C) \bullet \boldsymbol{\nu} = -\nabla\boldsymbol{\nu} \bullet \nabla\gamma^C \qquad \textbf{(6.27f)}$$

where the notation is the same as that used following equations (6.1a) through (6.1f).

The proofs of the identities in equations (6.27a) through (6.27f) are similar. As an example, consider the left side of equation (6.27a). Application of the chain rule gives:

$$(\nabla\nabla\gamma)\bullet\boldsymbol{\nu} = \nabla(\boldsymbol{\nu}\bullet\nabla\gamma) - \nabla\boldsymbol{\nu}\bullet\nabla\gamma \tag{6.28}$$

Because $\nabla\gamma$ is a vector tangent to $\mathbf{n}$, $\boldsymbol{\nu}\bullet\nabla\gamma = 0$ and the first term on the right side of equation (6.28) is zero. Thus equation (6.27a) results.

Some other second derivatives of generalized functions that arise are those involving a mix of spatial, surficial, and curvilineal del operators. The following identities apply to these terms:

Identity 6.8a

$$\nabla^s\nabla\gamma = \mathbf{n}\bullet\nabla\gamma(\nabla\mathbf{n})^T \tag{6.29a}$$

Identity 6.8b

$$\nabla^c\nabla\gamma = -\boldsymbol{\lambda}\nabla\boldsymbol{\lambda}\bullet\nabla\gamma \tag{6.29b}$$

Identity 6.8c

$$\nabla^s\nabla\gamma^s = \nabla\nabla\gamma^s + \mathbf{n}\nabla\mathbf{n}\bullet\nabla\gamma^s \tag{6.29c}$$

Identity 6.8d

$$\nabla^c\nabla\gamma^s = -\boldsymbol{\lambda}\nabla\boldsymbol{\lambda}\bullet\nabla\gamma^s \tag{6.29d}$$

Identity 6.8e

$$\nabla^s\nabla\gamma^c = \nabla\nabla\gamma^c + \mathbf{n}\nabla\mathbf{n}\bullet\nabla\gamma^c \tag{6.29e}$$

Identity 6.8f

$$\nabla^c\nabla\gamma^c = \nabla\nabla\gamma^c - \boldsymbol{\lambda}\bullet\nabla\gamma^c(\nabla\boldsymbol{\lambda})^T \tag{6.29f}$$

where the superscript T indicates the transpose of the tensor expression and the notation used is described following equations (6.1b), (6.1d), and (6.1f).

Note that the order of the del operators in these identities may not be simply interchanged as the tensor expressions are not symmetric. The proofs of the identities follow directly from equations (6.27a) through (6.27f) and are all similar in format, although some are a bit more tedious than others. As a relatively simple example, consider the proof of equation (6.29d). By definition of the curvilineal del operator, $\nabla^c = \boldsymbol{\lambda}\boldsymbol{\lambda}\bullet\nabla$, the left side of the equation may be written such that:

$$\nabla^c \nabla \gamma^s = \boldsymbol{\lambda}(\boldsymbol{\lambda} \bullet \nabla\nabla \gamma^s) \tag{6.30}$$

where the parentheses are employed for emphasis. Next, the term in parentheses may be replaced using Identity 6.7d, equation (6.27d), such that Identity 6.8d follows directly.

In the subsequent sections of this chapter, a few selected integration and averaging theorems will be proven in detail that rely on the identities involving generalized functions developed in this section. In Chapters 7 and 8, additional theorems, proved using the same methodology, will simply be tabulated. However, before proceeding to the example derivations of the integration and averaging theorems, three useful relations involving derivatives of unit vectors will be presented.

6.2 IDENTITIES INVOLVING DERIVATIVES OF UNIT VECTORS

In the derivation of theorems to be presented here and for use in obtaining other formulas, several identities involving the time and space derivatives of a unit vector will prove useful. These identities will be collected here along with their proofs. Although the theorems are written in terms of the unit vector $\mathbf{n}$, they apply to any of the unit vectors with suitable permutation of the coordinate directions.

6.2.1 Proof that $\nabla \bullet \mathbf{n} = \nabla^s \bullet \mathbf{n}$

To obtain the proof of this relation, apply equation (2.26) that defines the surface divergence in terms of the spatial divergence, $\nabla^s = \nabla - \mathbf{nn} \bullet \nabla$ to obtain:

$$\nabla \bullet \mathbf{n} = \nabla^s \bullet \mathbf{n} + (\mathbf{nn} \bullet \nabla) \bullet \mathbf{n} \tag{6.31}$$

Expansion of the last term in this equation yields:

$$\nabla \bullet \mathbf{n} = \nabla^s \bullet \mathbf{n} + \mathbf{n} \bullet \nabla \mathbf{n} \bullet \mathbf{n} \tag{6.32}$$

Because $\nabla \mathbf{n} \bullet \mathbf{n} = \nabla(\mathbf{n} \bullet \mathbf{n})/2 = 0$, the last term in equation (6.32) is zero and the desired identity is proven:

Identity 6.9

$$\nabla \bullet \mathbf{n} = \nabla^s \bullet \mathbf{n} \tag{6.33}$$

6.2.2 Proof that $\nabla^s \times \mathbf{n} = \mathbf{0}$

This proof makes use of the fact that a cross product between vectors will be orthogonal to those vectors so that $\nabla^s \times \mathbf{n}$ will be orthogonal to $\mathbf{n}$ such that:

$$(\nabla^s \times \mathbf{n}) \bullet \mathbf{n} = 0 \tag{6.34}$$

The operator ∇^s is surficial operator in the surface with $\mathbf{n}$ as the unit normal such that $\nabla^s = \boldsymbol{\lambda\lambda} \bullet \nabla + \boldsymbol{\nu\nu} \bullet \nabla$ and the following identity holds:

$$\nabla^s \times \mathbf{n} = (\boldsymbol{\lambda\lambda} \bullet \nabla + \boldsymbol{\nu\nu} \bullet \nabla) \times \mathbf{n} \tag{6.35}$$

Distribution of the terms of the right side yields:

$$\nabla^s \times \mathbf{n} = -(\boldsymbol{\lambda} \bullet \nabla \mathbf{n}) \times \boldsymbol{\lambda} - (\boldsymbol{\nu} \bullet \nabla \mathbf{n}) \times \boldsymbol{\nu} \tag{6.36}$$

Equations (2.10a) and (2.10b) provide, respectively, the relations among orthogonal unit vectors, $\boldsymbol{\lambda} \times \mathbf{n} = \boldsymbol{\nu}$ and $\mathbf{n} \times \boldsymbol{\nu} = \boldsymbol{\lambda}$, that may be substituted into equation (6.36) to obtain:

$$\nabla^s \times \mathbf{n} = -(\boldsymbol{\lambda} \bullet \nabla \mathbf{n}) \times (\mathbf{n} \times \boldsymbol{\nu}) - (\boldsymbol{\nu} \bullet \nabla \mathbf{n}) \times (\boldsymbol{\lambda} \times \mathbf{n}) \tag{6.37}$$

The two terms on the right side of this equation each involve a double cross product. Such a term satisfies the well-known identity [e.g. *Zill and Cullen*, 1992]:

$$\mathbf{a} \times (\mathbf{b} \times \mathbf{c}) = (\mathbf{a} \bullet \mathbf{c})\, \mathbf{b} - (\mathbf{a} \bullet \mathbf{b})\, \mathbf{c} \tag{6.38}$$

such that:

$$\nabla^s \times \mathbf{n} = -(\boldsymbol{\lambda} \bullet \nabla \mathbf{n} \bullet \boldsymbol{\nu})\, \mathbf{n} + (\boldsymbol{\lambda} \bullet \nabla \mathbf{n} \bullet \mathbf{n})\, \boldsymbol{\nu} - (\boldsymbol{\nu} \bullet \nabla \mathbf{n} \bullet \mathbf{n})\, \boldsymbol{\lambda} + (\boldsymbol{\nu} \bullet \nabla \mathbf{n} \bullet \boldsymbol{\lambda})\, \mathbf{n} \tag{6.39}$$

Because $\nabla \mathbf{n} \bullet \mathbf{n} = 0$, this equation reduces to:

$$\nabla^s \times \mathbf{n} = -(\boldsymbol{\lambda} \bullet \nabla \mathbf{n} \bullet \boldsymbol{\nu} - \boldsymbol{\nu} \bullet \nabla \mathbf{n} \bullet \boldsymbol{\lambda})\, \mathbf{n} \tag{6.40}$$

By equation (6.34), the right side of equation (6.40) must be orthogonal to $\mathbf{n}$. Therefore the term in parentheses must be zero and the desired identity is obtained:

Identity 6.10

$$\nabla^s \times \mathbf{n} = 0 \tag{6.41}$$

6.2.3 Proof that $d\mathbf{n}/dt = -\nabla^s \mathbf{w} \bullet \mathbf{n}$

The proof of this identity is more difficult than the previous two presented in this section and is, perhaps, best accomplished in the context of generalized functions. Thus the proof of this identity serves both as an exercise in using some of the identities for generalized functions developed previously and as an introduction to the techniques that must be used in proving the subsequent theorems of this manuscript.

Consider **n** to be a unit normal vector to some surface, S, in space. The integral of the partial time derivative in space of **n** over the surface may be transformed to an integral over an infinite volume by using the surficial generalized function, γ^s, and the spatial generalized function, γ, as in equation (4.8) to obtain:

$$\int_S \left.\frac{\partial \mathbf{n}}{\partial t}\right|_{\mathbf{x}} dS = -\int_{V_\infty} \left.\frac{\partial \mathbf{n}}{\partial t}\right|_{\mathbf{x}} \gamma^s \mathbf{n} \bullet \nabla \gamma dV \tag{6.42}$$

Application of the chain rule to the right side of this expression yields:

$$\int_S \left.\frac{\partial \mathbf{n}}{\partial t}\right|_{\mathbf{x}} dS = -\int_{V_\infty} \left.\frac{\partial (\mathbf{nn} \bullet \nabla \gamma)}{\partial t}\right|_{\mathbf{x}} \gamma^s dV + \int_{V_\infty} \mathbf{n} \left.\frac{\partial (\mathbf{n} \bullet \nabla \gamma)}{\partial t}\right|_{\mathbf{x}} \gamma^s dV \tag{6.43}$$

Because $\nabla \gamma$ is a vector in the **n** direction, $\mathbf{nn} \bullet \nabla \gamma \equiv \nabla \gamma$. Furthermore, equation (6.1a) indicates that $(\nabla \gamma) \bullet \partial \mathbf{n}/\partial t = 0$. Therefore, expansion of the derivatives in equation (6.43) and rearrangement leads to:

$$\int_S \left.\frac{\partial \mathbf{n}}{\partial t}\right|_{\mathbf{x}} dS = -\int_{V_\infty} \nabla^s \left(\left.\frac{\partial \gamma}{\partial t}\right|_{\mathbf{x}}\right) \gamma^s dV \tag{6.44}$$

where, again, $\nabla^s = \nabla - \mathbf{nn} \bullet \nabla$. Equation (3.42) provides the information that $\partial \gamma / \partial t = -\mathbf{w} \bullet \nabla \gamma$ where **w** is the velocity of the surface. Substitution of this expression into equation (6.44) and application of the surface gradient operator yields:

$$\int_S \left.\frac{\partial \mathbf{n}}{\partial t}\right|_{\mathbf{x}} dS = \int_{V_\infty} (\nabla^s \mathbf{w}) \bullet (\nabla \gamma) \gamma^s dV + \int_{V_\infty} \nabla^s (\nabla \gamma) \bullet \mathbf{w} \gamma^s dV \tag{6.45}$$

Identity 6.8a may be substituted into the second integral on the right side of this equation to obtain:

$$\int_S \left.\frac{\partial \mathbf{n}}{\partial t}\right|_{\mathbf{x}} dS = \int_{V_\infty} (\nabla^s \mathbf{w}) \bullet (\nabla \gamma) \gamma^s dV + \int_{V_\infty} \mathbf{w} \bullet \nabla \mathbf{n} (\mathbf{n} \bullet \nabla \gamma) \gamma^s dV \tag{6.46}$$

Now the volume integrals may be converted back to surface integrals using equation (4.8) so that the resulting equality is:

$$\int_S \left.\frac{\partial \mathbf{n}}{\partial t}\right|_{\mathbf{x}} dS = -\int_S (\nabla^s \mathbf{w}) \bullet \mathbf{n} dS - \int_S \mathbf{w} \bullet \nabla \mathbf{n} dS \qquad \textbf{(6.47)}$$

Because the surface is arbitrary, equation (6.47) must apply for every point on the surface. Thus this equation may be written at the microscale as:

Identity 6.11a

$$\left.\frac{\partial \mathbf{n}}{\partial t}\right|_{\mathbf{x}} = -(\nabla^s \mathbf{w}) \bullet \mathbf{n} - \mathbf{w} \bullet \nabla \mathbf{n} \qquad \textbf{(6.48a)}$$

or as:

Identity 6.11b

$$\frac{d\mathbf{n}}{dt} = -(\nabla^s \mathbf{w}) \bullet \mathbf{n} \qquad \textbf{(6.48b)}$$

where the total derivative is defined by:

$$\frac{d\mathbf{n}}{dt} = \left.\frac{\partial \mathbf{n}}{\partial t}\right|_{\mathbf{x}} + \mathbf{w} \bullet \nabla \mathbf{n} \qquad \textbf{(6.49)}$$

Relation (6.48b) follows also from a more classical analysis based on obtaining an expression for the material derivative of a surface element [see, e.g. *Batchelor* 1967; *Eringen*, 1980; *Stone*, 1990].

6.3 DERIVATION OF THE DIVERGENCE THEOREM *D*[3, (0, 0), 3]

Consider a single phase system with volume V that is bounded by the closed surface S. Define a spatial generalized function $\gamma(\mathbf{x})$ such that $\gamma = 1$ inside V and $\gamma = 0$ outside V. The integral of the divergence of a vector field $\mathbf{f}(\mathbf{x})$ over the volume V is then given by:

$$\int_V \nabla \bullet \mathbf{f} dV = \int_{V_\infty} (\nabla \bullet \mathbf{f}) \gamma dV \qquad \textbf{(6.50)}$$

Note that in equation (6.50), the differentiation and the integration are carried out in the same coordinate system. Application of the chain rule to the right side of equation (6.50) yields:

$$\int_V \nabla \bullet \mathbf{f} dV = \int_{V_\infty} \nabla \bullet (\mathbf{f}\gamma) dV - \int_{V_\infty} (\nabla \gamma) \bullet \mathbf{f} dV \qquad \textbf{(6.51)}$$

The first term of the right hand side is zero according to equation (6.20b) so that equation (6.51) reduces to:

$$\int_V \nabla \bullet \mathbf{f} dV = -\int_{V_\infty} (\nabla \gamma) \bullet \mathbf{f} dV \tag{6.52}$$

The integral over V_∞ is changed to a surface integral using equation (4.8) to obtain:

D[3, (0, 0), 3]

$$\int_V \nabla \bullet \mathbf{f} dV = \int_S \mathbf{n}^* \bullet \mathbf{f} dS \tag{6.53}$$

where $\mathbf{n}^*$ is the unit vector normal to S that is positive outward from V. Equation (6.53) is the well known divergence theorem. Comparison of the derivation given here with a more classical approach [e.g. *Kreyszig*, 1962; *Slattery*, 1972; *Whitaker*, 1968; *Zill and Cullen*, 1992] shows the relative simplicity of the derivation using the spatial generalized function γ.

6.4 DERIVATION OF AVERAGING THEOREMS *G, D, C,* AND *T*[3, (1, 0), 2]

6.4.1 Specification of the Averaging Volume

Consider a single or multiphase system for which spatially averaged one-dimensional equations must be developed, i.e. the equations must be formulated in one macroscopic dimension and two megascopic dimensions. An example of such a system is the flow in a groundwater aquifer with a length much larger than the width and the height. A single phase system requiring a similar analysis would be the flow in a river in the direction of the axis of the river channel. The averaging volume or REV is defined to be a slab of constant thickness, bounded by two plane cross-sections perpendicular to the length axis of the system, and the boundary of the system itself as in figure 4.10. The thickness of the slab is chosen so that equation (4.1) is satisfied requiring that the microscale be much smaller than the macroscale which, in turn, must be much smaller than the megascale. Additionally, the requirement that both faces of the REV be orthogonal to the axis of the system supplements the requirements on relative length scales by implying that the change in direction of the axis is negligible over the slab thickness. The volume of the REV is denoted as V while the fraction of the REV occupied by the α-phase is denoted as V_α. The system under study is located in global space indicated as V_∞.

The local coordinate system in the REV is indicated by $\boldsymbol{\xi}$ and is defined with respect to the macroscopic position of the REV. The coordinate location of the REV in the global coordinate system is indicated by $\mathbf{x}$. Although the plane cross-sections bounding an REV are fixed in time, the REV itself is time dependent due to a possible movement of the system boundaries. The thickness of the REV does not vary with position although the orientation of the REV may depend on position. For the derivation of the averaging theorems, it is convenient to define the macroscopic coordinate N normal to the face of the averaging slab and the microscopic coordinates $\mathbf{u}_\xi$ in the directions tangent to the face of the slab. With n the microscopic coordinate in the direction normal to the face of the slab, a position in space is completely defined by $(N+n, \mathbf{u}_\xi)$. A unit vector $\mathbf{N}$ is also defined that is normal to the face of the slab and positive in direction N. The curvature of the axis at the macroscale must be such that:

$$D|\mathbf{N}\bullet\nabla\mathbf{N}| \ll 1 \tag{6.54}$$

To derive the averaging theorems for the α-phase in a multiphase system, two generalized functions will be defined:

$$\gamma^\alpha(\mathbf{x}, \boldsymbol{\xi}, t) = \gamma^\alpha(N+n, \mathbf{u}_\xi, t) = \begin{cases} 1 & \text{in the } \alpha\text{-phase} \\ 0 & \text{elsewhere in space} \end{cases} \tag{6.55a}$$

and:

$$\gamma(N, \boldsymbol{\xi}, t) = \begin{cases} 1 & \text{at all points in an REV} \\ 0 & \text{at all points outside the REV} \end{cases} \tag{6.55b}$$

Note, that γ is a function of both the macroscopic coordinate N and the microscopic coordinates $\boldsymbol{\xi}$ because the size and shape of the REV may vary with the position. Note also that both γ^α and γ are time dependent.

6.4.2 Derivation of *G*[3, (1, 0), 2]

This derivation involves integration over the averaging slab of the gradient of a spatial scalar function $f(\mathbf{x}, t)$. Consistent with the functional dependence of the generalized functions explained in the last Subsection, the function f may be expressed as $f(N+n, \mathbf{u}_\xi, t)$ where n is the microscopic coordinate in the direction normal to the slab (i.e. in direction N), and $\mathbf{u}_\xi$ are the microscopic surface coordinates tangent to the slab (i.e. normal to N). For derivation of this theorem, the gradient operator is expressed as:

$$\nabla = \nabla^C + \nabla^s \tag{6.56}$$

where $\nabla^C = \mathbf{NN}\bullet\nabla$ and involves differentiation with respect to the macroscopic coordinate while ∇^s indicates the two-dimensional gradient with respect to the $\mathbf{u}_\xi$ coordinates.

The integral over the α-phase of the gradient of f can then be written as:

$$\int_{V_\alpha} \nabla f dV = \int_{V_\alpha} \nabla^C f dV + \int_{V_\alpha} \nabla^s f dV \tag{6.57}$$

where the integration is over the microscopic coordinates. Transformation of the limits of integration to integration over all space is accomplished by multiplying the integrands by the generalized functions γ^α and γ defined in equations (6.55a) and (6.55b), respectively. Therefore equation (6.57) becomes:

$$\int_{V_\alpha} \nabla f dV = \int_{V_\infty} (\nabla^C f)\, \gamma\gamma^\alpha dV + \int_{V_\infty} (\nabla^s f)\, \gamma\gamma^\alpha dV \tag{6.58}$$

Application of the chain rule to the two integrals on the right side of this equation yields:

$$\int_{V_\alpha} \nabla f dV = \int_{V_\infty} \nabla^C (f\gamma\gamma^\alpha)\, dV - \int_{V_\infty} f\nabla^C (\gamma\gamma^\alpha)\, dV$$

$$+ \int_{V_\infty} \nabla^s (f\gamma\gamma^\alpha)\, dV - \int_{V_\infty} f\nabla^s (\gamma\gamma^\alpha)\, dV \tag{6.59}$$

The order of integration and differentiation may be directly interchanged in the first term on the right side since the macroscopic variable of differentiation is not a variable of integration. Then the generalized functions in this term may be eliminated if the integration is restated as being over V_α. The third integral on the right side of equation (6.59) is zero because integration results in evaluation of terms at the edge of the V_∞ averaging slab where $\gamma = 0$. With these manipulations applied, and with some rearrangement of the remaining terms, equation (6.59) becomes:

$$\int_{V_\alpha} \nabla f dV = \nabla^C \int_{V_\alpha} f dV - \int_{V_\infty} [(\nabla^C + \nabla^s)\, \gamma^\alpha]\, \gamma f dV - \int_{V_\infty} [(\nabla^C + \nabla^s)\, \gamma]\, \gamma^\alpha f dV \tag{6.60}$$

The gradients of γ and γ^α that appear in this equation can be treated using the procedures of Section 3.4. First, note that γ^α is a function of $N + n$ so the differentiation of γ^{α} with respect to the macroscopic coordinate in the direction normal to the faces of the averaging volume is identical to differentiation with respect to the microscopic coordinate in that direction. Thus, by equation (3.37):

$$(\nabla^C + \nabla^s)\,\gamma^{\alpha} = \nabla\gamma^{\alpha} = -\mathbf{n}\,\delta[g(N+n, \mathbf{u}_{\xi};\mathbf{x}^{\alpha\beta}(t))] \qquad \textbf{(6.61)}$$

where $\mathbf{x}^{\alpha\beta}(t)$ is the location of the interface between the α-phase and all other phases, and $\mathbf{n}$ is the unit normal vector on these interfaces pointing outward from the α-phase. If this expression is used in the second term on the right side of equation (6.60) to convert it to a surface integral, then:

$$\int_{V_\alpha} \nabla f dV = \nabla^C \int_{V_\alpha} f dV + \int_{S_{\alpha\beta}} \mathbf{n} f dS - \int_{V_\infty} [(\nabla^C + \nabla^s)\,\gamma]\,\gamma^{\alpha} f dV \qquad \textbf{(6.62)}$$

In the second integral on the right side, γ was eliminated such that the region of integration was restricted to the interface between the α-phase and all other phases within the averaging volume. The term $\nabla^C\gamma$ in the third integral on the right side of equation (6.62) gives rise to contributions only along the edge of the physical region of interest. Because the thickness of the averaging slab is the same for each averaging volume, the derivative of γ in the direction normal to the face of the slab will be zero when taken with the microscopic coordinate held constant. Therefore:

$$(\nabla^C + \nabla^s)\,\gamma = -\mathbf{n}^*\,\delta[g(N, \xi;\mathbf{x}^{\text{edge}}(t))] \qquad \textbf{(6.63)}$$

where $\mathbf{n}^*$ is the unit normal vector on the system boundary pointing outward and $\mathbf{x}^{\text{edge}}(t)$ is the location of the edge of the averaging volume. Substitution of equation (6.63) into equation (6.62) finally gives:

G[3, (1, 0), 2]

$$\int_{V_\alpha} \nabla f dV = \nabla^C \int_{V_\alpha} f dV + \int_{S_{\alpha\beta}} \mathbf{n} f dS + \int_{S_{\alpha_{\text{edge}}}} \mathbf{n}^* f dS \qquad \textbf{(6.64)}$$

In this equation, integration is only over the α-portion of the edge surface because of the presence of γ^{α} in the last term in equation (6.62).

6.4.3 Derivation of *D*[3, (1, 0), 2]

This derivation involves integration over the averaging slab of the divergence of a vector field $\mathbf{f}(\mathbf{x}, t)$. Identically to the specification used in the derivation of $G[3, (1, 0), 2]$, the del operator for the divergence of a spatial vector function is expressed as a combination of a component in the macroscopic direction normal to the face of the averaging slab and microscopic components tangent to the slab face as in equation (6.56). Then, based on equation (2.49), since **f** may be expressed as a function of the combination of microscopic and macroscopic coordinates, $\mathbf{f}(N+n, \mathbf{u}_\xi, t)$, the divergence of **f** is given by:

$$\nabla\bullet\mathbf{f} = \nabla^C\bullet\mathbf{f}^C + \nabla^s\bullet\mathbf{f}^s - \mathbf{N}\bullet\nabla\mathbf{N}\bullet\mathbf{f} + \mathbf{N}\bullet\mathbf{f}\nabla\bullet\mathbf{N} \tag{6.65}$$

Recall that ∇^C is differentiation with respect to macroscopic coordinates and ∇^s indicates surface divergence with respect to microscopic coordinates. The integral under study is therefore:

$$\int_{V_\alpha}\nabla\bullet\mathbf{f}dV = \int_{V_\alpha}\nabla^C\bullet\mathbf{f}^C dV + \int_{V_\alpha}\nabla^s\bullet\mathbf{f}^s dV - \int_{V_\alpha}\mathbf{N}\bullet\nabla\mathbf{N}\bullet\mathbf{f}dV + \int_{V_\alpha}\mathbf{N}\bullet\mathbf{f}\nabla\bullet\mathbf{N}dV \tag{6.66}$$

Because $\mathbf{N} = \mathbf{N}(N)$ is a macroscopic function and does not depend on microscopic coordinates, $\nabla\bullet\mathbf{N} = 0$. Therefore the last term in equation (6.66) is zero. The generalized functions for the REV and for the α-phase may now be employed such that:

$$\int_{V_\alpha}\nabla\bullet\mathbf{f}dV = \int_{V_\infty}(\nabla^C\bullet\mathbf{f}^C)\,\gamma\gamma^\alpha dV + \int_{V_\infty}(\nabla^s\bullet\mathbf{f}^s)\,\gamma\gamma^\alpha dV - \int_{V_\alpha}\mathbf{N}\bullet\nabla\mathbf{N}\bullet\mathbf{f}dV \tag{6.67}$$

Application of the chain rule to the first two integrals on the right side of this equation and collection of some of the resulting terms yields:

$$\int_{V_\alpha}\nabla\bullet\mathbf{f}dV = \int_{V_\infty}\nabla^C\bullet(\mathbf{f}^C\gamma\gamma^\alpha)\,dV + \int_{V_\infty}\nabla^s\bullet(\mathbf{f}^s\gamma\gamma^\alpha)\,dV$$

$$-\int_{V_\infty}[\nabla^C(\gamma\gamma^\alpha)]\bullet\mathbf{f}^C dV - \int_{V_\infty}[\nabla^s(\gamma\gamma^\alpha)]\bullet\mathbf{f}^s dV - \int_{V_\alpha}\mathbf{N}\bullet\nabla\mathbf{N}\bullet\mathbf{f}dV \tag{6.68}$$

The first four terms on the right side of this equation are treated in much the same way as in the development following equation (6.59). The order of integration and differentiation in the first term is interchanged since the macroscopic variable of differentiation is not a variable of integration, and the domain of integration is then changed back to V_α by eliminating the generalized functions from the integrand. The second integral on the right side of equation (6.68) is zero because integration results in evaluation of terms at the edge of the averaging slab where $\gamma = 0$. Note also that $\mathbf{f}^C$ and $\mathbf{f}^s$ in the third and fourth integrals may be replaced by $\mathbf{f}$ because the dot product with ∇^C and ∇^s, respectively will preserve the value of the integrand. With these manipulations applied, and with some rearrangement, equation (6.68) becomes:

$$\int_{V_\alpha} \nabla \cdot \mathbf{f} dV = \nabla^C \cdot \int_{V_\alpha} \mathbf{f}^C dV - \int_{V_\infty} [(\nabla^C + \nabla^s)\gamma^\alpha] \cdot \mathbf{f}\gamma dV$$

$$- \int_{V_\infty} [(\nabla^C + \nabla^s)\gamma] \cdot \mathbf{f}\gamma^\alpha dV - \int_{V_\alpha} \mathbf{N} \cdot \nabla \mathbf{N} \cdot \mathbf{f} dV \qquad \textbf{(6.69)}$$

The gradients of γ^α and γ that appear in this equation can be treated using equations (6.61) and (6.63), respectively. The entry of the delta function into the integrals then converts the volume integrals to surface integrals and equation (6.69) becomes:

D[3, (1, 0), 2]

$$\int_{V_\alpha} \nabla \cdot \mathbf{f} dV = \nabla^C \cdot \int_{V_\alpha} \mathbf{f}^C dV + \int_{S_{\alpha\beta}} \mathbf{n} \cdot \mathbf{f} dV + \int_{S_{\alpha_{\text{edge}}}} \mathbf{n}^* \cdot \mathbf{f} dV - \int_{V_\alpha} \mathbf{N} \cdot \nabla \mathbf{N} \cdot \mathbf{f} dV$$

(6.70)

6.4.4 Derivation of *C*[3, (1, 0), 2]

Consider the integration of the spatial curl of a spatial vector field, $\mathbf{f}(\mathbf{x}, t)$, expressed for this derivation as $\mathbf{f}(N + n, \mathbf{u}_\xi, t)$, over the averaging slab. The generalized functions γ and γ^α defined in Subsection 6.4.1 can be used to convert the integral over V_α to an integral over V_∞:

$$\int_{V_\alpha} \nabla \times \mathbf{f} dV = \int_{V_\infty} (\nabla \times \mathbf{f})\gamma\gamma^\alpha dV \qquad \textbf{(6.71)}$$

Application of the chain rule to the right side of equation (6.71) yields:

$$\int_{V_\alpha} \nabla \times \mathbf{f} dV = \int_{V_\infty} \nabla \times (\mathbf{f}\gamma\gamma^\alpha) dV - \int_{V_\infty} (\nabla \gamma^\alpha) \times \mathbf{f}\gamma dV - \int_{V_\infty} (\nabla \gamma) \times \mathbf{f}\gamma^\alpha dV$$

(6.72)

The last two terms in this expression are converted to surface integrals over the $\alpha\beta$-interface and the edge of the region of interest, respectively, by the same reasoning using the generalized functions as employed in the preceding sections. Thus equation (6.72) becomes:

$$\int_{V_\alpha} \nabla \times \mathbf{f} dV = \int_{V_\infty} \nabla \times (\mathbf{f}\gamma\gamma^\alpha)\, dV + \int_{S_{\alpha\beta}} \mathbf{n} \times \mathbf{f} dS + \int_{S_{\alpha_{\text{edge}}}} \mathbf{n}^* \times \mathbf{f} dS \qquad \textbf{(6.73)}$$

Further manipulations will involve only the first term on the right side.

The del operator may be decomposed into a macroscopic part and a microscopic part as in equation (6.56). The vector field $\mathbf{f}$ can be split in a similar way by letting:

$$\mathbf{f} = \mathbf{f}^C + \mathbf{f}^s \qquad \textbf{(6.74)}$$

where $\mathbf{f}^C = \mathbf{N}\mathbf{N}\bullet\mathbf{f}$ and $\mathbf{f}^s$ includes the components of $\mathbf{f}$ tangent to the slab. Using equations (6.56) and (6.74) in the first integral on the right side of equation (6.73), one obtains:

$$\int_{V_\infty} \nabla \times (\mathbf{f}\gamma\gamma^\alpha)\, dV = \int_{V_\infty} \nabla^C \times (\mathbf{f}^s\gamma\gamma^\alpha)\, dV + \int_{V_\infty} \nabla^C \times (\mathbf{f}^C\gamma\gamma^\alpha)\, dV$$

$$+ \int_{V_\infty} \nabla^s \times (\mathbf{f}\gamma\gamma^\alpha)\, dV \qquad \textbf{(6.75)}$$

In the first term on the right, the operator ∇^C may be moved outside of the integral and the generalized functions eliminated to return the domain of integration to V_α. The differential operator may also be moved outside the second integral, if desired; but alternatively equation (2.57) may be employed to obtain:

$$\nabla^C \times (\mathbf{f}^C\gamma\gamma^\alpha) = -(\mathbf{N}\bullet\nabla\mathbf{N}) \times \mathbf{f}^C\gamma\gamma^\alpha \qquad \textbf{(6.76)}$$

This second option is selected here and then the generalized functions are removed so that the domain of integration becomes V_α. The last integral in equation (6.75) is zero because performing the integration results in evaluation of the integrand at the "boundary" of V_∞ where γ is zero. Thus equation (6.75) becomes:

$$\int_{V_\infty} \nabla \times (\mathbf{f}\gamma\gamma^\alpha)\, dV = \nabla^C \times \int_{V_\alpha} \mathbf{f}^s dV - \int_{V_\alpha} (\mathbf{N}\bullet\nabla\mathbf{N}) \times \mathbf{f}^C dV \qquad \textbf{(6.77)}$$

Finally, substitution of this expression into equation (6.73) results in the theorem of interest:

C[3, (1, 0), 2]

$$\int_{V_\alpha} \nabla \times \mathbf{f} dV = \nabla^C \times \int_{V_\alpha} \mathbf{f}^s dV - \int_{V_\alpha} (\mathbf{N} \bullet \nabla \mathbf{N}) \times \mathbf{f}^C dV + \int_{S_{\alpha\beta}} \mathbf{n} \times \mathbf{f} dS + \int_{S_{\alpha_{\text{edge}}}} \mathbf{n}^* \times \mathbf{f} dS$$

(6.78)

6.4.5 Derivation of *T*[3, (1, 0), 2]

The integral over the α-phase in an REV of the time derivative of a spatial function f can be written in terms of the integral over the REV by making use of the generalized function γ^α as:

$$\int_{V_\alpha} \left.\frac{\partial f}{\partial t}\right|_{\mathbf{x}} dV = \int_{V} \left.\frac{\partial f}{\partial t}\right|_{\mathbf{x}} \gamma^\alpha dV \qquad \textbf{(6.79)}$$

The integral on the right side of this equation can be further changed to an integral over all space by using the generalized function γ to obtain:

$$\int_{V_\alpha} \left.\frac{\partial f}{\partial t}\right|_{\mathbf{x}} dV = \int_{V_\infty} \left.\frac{\partial f}{\partial t}\right|_{\mathbf{x}} \gamma^\alpha \gamma dV \qquad \textbf{(6.80)}$$

The time derivative is the partial derivative of f evaluated at a fixed position in space. Application of the chain rule to the right hand size of equation (6.80) gives:

$$\int_{V_\alpha} \left.\frac{\partial f}{\partial t}\right|_{\mathbf{x}} dV = \int_{V_\infty} \left.\frac{\partial (f\gamma^\alpha\gamma)}{\partial t}\right|_{\mathbf{x}} dV - \int_{V_\infty} f \left.\frac{\partial \gamma^\alpha}{\partial t}\right|_{\mathbf{x}} \gamma dV - \int_{V_\infty} f\gamma^\alpha \left.\frac{\partial \gamma}{\partial t}\right|_{\mathbf{x}} dV \qquad \textbf{(6.81)}$$

Because the boundaries of V_∞ are independent of time, differentiation and integration can be interchanged in the first term on the right side of equation (6.81). Then the definitions of the generalized functions may be used to convert the derivative of an integral over V_∞ to the derivative of an integral over V_α:

$$\int_{V_\infty} \left.\frac{\partial (f\gamma^\alpha\gamma)}{\partial t}\right|_{\mathbf{x}} dV = \frac{\partial}{\partial t}\int_{V_\infty} f\gamma^\alpha \gamma dV = \frac{\partial}{\partial t}\int_{V} f\gamma^\alpha dV = \frac{\partial}{\partial t}\int_{V_\alpha} f dV \qquad \textbf{(6.82)}$$

Note that the time derivative in the right side of this equation is evaluated holding the position of the REV fixed. Substitution of equation (6.82) into equation (6.81) yields:

$$\int_{V_\alpha} \frac{\partial f}{\partial t}\bigg|_{\mathbf{x}} dV = \frac{\partial}{\partial t}\int_{V_\alpha} f dV - \int_{V_\infty} f \frac{\partial \gamma^\alpha}{\partial t}\bigg|_{\mathbf{x}} \gamma dV - \int_{V_\infty} f\gamma^\alpha \frac{\partial \gamma}{\partial t}\bigg|_{\mathbf{x}} dV \qquad \textbf{(6.83)}$$

Equation (3.42) indicates that $\partial\gamma/\partial t|_{\mathbf{x}} = -\mathbf{w}\bullet\nabla\gamma$ where $\mathbf{w}$ is the velocity of the boundary of the REV. Similarly, the generalized function identifying the α-phase satisfies the identity $\partial\gamma^\alpha/\partial t|_{\mathbf{x}} = -\mathbf{w}\bullet\nabla\gamma^\alpha$ where $\mathbf{w}$ is the velocity of the interface between the α-phase and all other phases. With these equalities applied, equation (6.83) becomes:

$$\int_{V_\alpha} \frac{\partial f}{\partial t}\bigg|_{\mathbf{x}} dV = \frac{\partial}{\partial t}\int_{V_\alpha} f dV + \int_{V_\infty} f(\mathbf{w}\bullet\nabla\gamma^\alpha)\,\gamma dV + \int_{V_\infty} f\gamma^\alpha(\mathbf{w}\bullet\nabla\gamma)\,dV \qquad \textbf{(6.84)}$$

Equation (4.8) relates integrals over V_∞ when the integrand contains the gradient of a volumetric generalized function to an integral over the boundary of the region identified by the generalized function. This identity allows equation (6.84) to be transformed to:

$$\int_{V_\alpha} \frac{\partial f}{\partial t}\bigg|_{\mathbf{x}} dV = \frac{\partial}{\partial t}\int_{V_\alpha} f dV - \int_{S_{\alpha\beta_\infty}} f\mathbf{w}\bullet\mathbf{n}\gamma dS - \int_{S} f\gamma^\alpha\mathbf{w}\bullet\mathbf{n}^* dS \qquad \textbf{(6.85)}$$

where:

- $\mathbf{n}$ is the unit normal vector to $\alpha\beta$-interfaces, positive outward from the α-phase,
- $S_{\alpha\beta_\infty}$ is the interface of the α-phase with all other phases in V_∞,
- $\mathbf{n}^*$ is the outward directed unit normal vector on the boundary of the REV, and
- S is the boundary of the REV.

Note that the spatial generalized function γ restricts the second integrand on the right side such that it is nonzero only on interfaces within the REV. This surface area is indicated as $S_{\alpha\beta}$ and γ may be dropped from the integrand if the integration is only over this area. The REV has been defined such that in the last integral, the velocity of the surface of the REV, $\mathbf{w}$, will be nonzero only over the edges of the REV, not over the planar surfaces. Furthermore, the presence of the generalized function γ^α in the integrand means that contributions arise only from the portion of the boundary that intersects the α-phase. This surface is indicated as $S_{\alpha_{edge}}$. Thus with the generalized functions eliminated from the integrands in equation (6.85) and the surfaces of integration specified explicitly, the desired theorem results:

T[3, (1, 0), 2]

$$\int_{V_\alpha} \left.\frac{\partial f}{\partial t}\right|_{\mathbf{x}} dV = \frac{\partial}{\partial t}\int_{V_\alpha} f dV - \int_{S_{\alpha\beta}} \mathbf{n}\cdot\mathbf{w} f dS - \int_{S_{\alpha_{\text{edge}}}} \mathbf{n}^*\cdot\mathbf{w} f dS \tag{6.86}$$

6.5 DERIVATION OF INTEGRATION THEOREM *T*[2, (0, 0), 1]

The integration theorem $T[2, (0, 0), 1]$ is a transport theorem providing a relation for the integral over a general curve located on a surface of the time derivative of a function defined on that surface. Note that the function may be either a spatial or a surficial function. The partial time derivative of the function considered is that taken keeping the surface coordinates constant. Equation (2.68) relates this time derivative to the time derivative keeping spatial coordinates constant:

$$\left.\frac{\partial f}{\partial t}\right|_{\mathbf{u}} = \left.\frac{\partial f}{\partial t}\right|_{\mathbf{x}} + \mathbf{w}\cdot\mathbf{nn}\cdot\nabla f \tag{6.87}$$

where $\mathbf{w}$ is the velocity of the surface and $\mathbf{n}$ is the unit vector normal to the surface. Thus the integral under consideration is:

$$\int_C \left.\frac{\partial f}{\partial t}\right|_{\mathbf{u}} dC = \int_C \left.\frac{\partial f}{\partial t}\right|_{\mathbf{x}} dC + \int_C \mathbf{w}\cdot\mathbf{nn}\cdot\nabla f dC \tag{6.88}$$

where C is the curve in the surface of interest. The theorem of interest will relate the integral of the derivative on the left side of equation (6.88) to the time derivative of the integral of f.

In order to obtain the desired relation, the change of integration region is facilitated using generalized functions and the techniques described in detail in Section 4.4.2. The transformation used to change an integral over a curve to an integral over all space is given in equation (4.11). Employing this relation in the first integral on the right side of equation (6.88), one obtains:

$$\int_C \left.\frac{\partial f}{\partial t}\right|_{\mathbf{u}} dC = \int_{V_\infty} \left.\frac{\partial f}{\partial t}\right|_{\mathbf{x}} \gamma^c (\boldsymbol{\nu}\cdot\nabla^s\gamma^s)\,(\mathbf{n}\cdot\nabla\gamma)\,dV + \int_C \mathbf{w}\cdot\mathbf{nn}\cdot\nabla f dC \tag{6.89}$$

where the notation used for the generalized functions and for the unit normal vectors is the same as given in Section 4.4.2. The subsequent manipulations will involve the integral over V_∞. Identity 6.3a, given by equation (6.12a), allows the surface gradient of γ^s to be replaced by the spatial gradient. The chain rule is now applied to the time derivative (where for convenience the stipulation that the time derivative is taken holding spatial coordinates constant is not explicitly noted) to obtain:

$$\int_{V_\infty} \frac{\partial f}{\partial t}\bigg|_{\mathbf{x}} \gamma^c (\boldsymbol{\nu}\bullet\nabla^s\gamma^s)(\mathbf{n}\bullet\nabla\gamma)\,dV = \int_{V_\infty} \frac{\partial\,[f\gamma^c(\boldsymbol{\nu}\bullet\nabla\gamma^s)(\mathbf{n}\bullet\nabla\gamma)]}{\partial t}\,dV$$

$$-\int_{V_\infty} f\frac{\partial\gamma^c}{\partial t}(\boldsymbol{\nu}\bullet\nabla\gamma^s)(\mathbf{n}\bullet\nabla\gamma)\,dV - \int_{V_\infty} f\gamma^c\left(\frac{\partial\boldsymbol{\nu}}{\partial t}\bullet\nabla\gamma^s\right)(\mathbf{n}\bullet\nabla\gamma)\,dV$$

$$-\int_{V_\infty} f\gamma^c\left(\boldsymbol{\nu}\bullet\nabla\frac{\partial\gamma^s}{\partial t}\right)(\mathbf{n}\bullet\nabla\gamma)\,dV - \int_{V_\infty} f\gamma^c(\boldsymbol{\nu}\bullet\nabla\gamma^s)\left(\frac{\partial\mathbf{n}}{\partial t}\bullet\nabla\gamma\right)dV$$

$$-\int_{V_\infty} f\gamma^c(\boldsymbol{\nu}\bullet\nabla\gamma^s)\left(\mathbf{n}\bullet\nabla\frac{\partial\gamma}{\partial t}\right)dV \qquad \textbf{(6.90)}$$

The task now is to simplify the six integrals on the right side of equation (6.90).

The order of integration and differentiation may be interchanged in the first integral because the limits of integration are independent of time. Then equation (4.11) may be invoked to convert the integral over V_∞ to an integral over C. In making this change, the partial derivative must be converted to a total derivative, indicating that the curve is followed as it moves in space. Thus the first term on the right in equation (6.90) rearranges to:

$$\int_{V_\infty} \frac{\partial\,[f\gamma^c(\boldsymbol{\nu}\bullet\nabla\gamma^s)(\mathbf{n}\bullet\nabla\gamma)]}{\partial t}\,dV = \frac{\partial}{\partial t}\int_{V_\infty} f\gamma^c(\boldsymbol{\nu}\bullet\nabla\gamma^s)(\mathbf{n}\bullet\nabla\gamma)\,dV = \frac{d}{dt}\int_C f\,dC$$

$$\textbf{(6.91a)}$$

The time derivative of the curvilineal generalized function γ^c in the second term on the right side of equation (6.90) can be written in terms of its gradient by invoking Identity 6.5b, given as equation (6.19b). Then working backwards by sequentially invoking equations (4.11), (4.10), and (3.26b) for a single curve, one obtains:

$$-\int_{V_\infty} f\frac{\partial\gamma^c}{\partial t}(\boldsymbol{\nu}\bullet\nabla\gamma^s)(\mathbf{n}\bullet\nabla\gamma)\,dV = \int_{V_\infty} f\mathbf{w}\bullet\nabla\gamma^c(\boldsymbol{\nu}\bullet\nabla\gamma^s)(\mathbf{n}\bullet\nabla\gamma)\,dV$$

$$= -\int_{S_\infty} f\mathbf{w}\bullet\nabla\gamma^c(\boldsymbol{\nu}\bullet\nabla\gamma^s)\,dS = \int_{C_\infty} f\mathbf{w}\bullet\nabla\gamma^c\,dC = -f\mathbf{w}\bullet\mathbf{e}\big|_{\text{ends}}$$

$$\textbf{(6.91b)}$$

where $\mathbf{e}$ is the unit vector tangent to and pointing outward from C at its ends.

The third and fifth integrals on the right side of equation (6.90) are zero by equations (6.1c) and (6.1a), respectively. Equations (6.19a) and (3.42) may be used, respectively, to relate the time derivative of γ^s in the fourth term and the time derivative of γ in the sixth term to their gradients. With the preceding changes implemented, equation (6.90) becomes:

$$\int_{V_\infty} \left.\frac{\partial f}{\partial t}\right|_{\mathbf{x}} \gamma^c (\boldsymbol{\nu}\bullet\nabla^s\gamma^s)(\mathbf{n}\bullet\nabla\gamma)\,dV = \frac{d}{dt}\int_C f dC - f\mathbf{w}\bullet\mathbf{e}|_{\text{ends}}$$

$$+\int_{V_\infty} f\gamma^c[\boldsymbol{\nu}\bullet\nabla(\mathbf{w}\bullet\nabla\gamma^s)](\mathbf{n}\bullet\nabla\gamma)\,dV + \int_{V_\infty} f\gamma^c(\boldsymbol{\nu}\bullet\nabla\gamma^s)[\mathbf{n}\bullet\nabla(\mathbf{w}\bullet\nabla\gamma)]\,dV$$

(6.92)

Further work is required on the third and fourth terms on the right side of this equation to convert them back to integrals over C. The second integral will be discussed in detail, with manipulations on the third integral being analogous.

The chain rule is applied to the integral to obtain:

$$\int_{V_\infty} f\gamma^c[\boldsymbol{\nu}\bullet\nabla(\mathbf{w}\bullet\nabla\gamma^s)](\mathbf{n}\bullet\nabla\gamma)\,dV = \int_{V_\infty}\nabla\bullet[\boldsymbol{\nu}f\gamma^c(\mathbf{w}\bullet\nabla\gamma^s)(\mathbf{n}\bullet\nabla\gamma)]\,dV$$

$$-\int_{V_\infty}\nabla\bullet[\boldsymbol{\nu}f\gamma^c(\mathbf{n}\bullet\nabla\gamma)](\mathbf{w}\bullet\nabla\gamma^s)\,dV \quad \textbf{(6.93)}$$

The first term on the right is zero by Identity 6.6b, equation (6.20b). Expansion of the second term making use of the chain rule re-expresses equation (6.93) as:

$$\int_{V_\infty} f\gamma^c[\boldsymbol{\nu}\bullet\nabla(\mathbf{w}\bullet\nabla\gamma^s)](\mathbf{n}\bullet\nabla\gamma)\,dV = -\int_{V_\infty}(\nabla\bullet\boldsymbol{\nu})f\gamma^c(\mathbf{n}\bullet\nabla\gamma)(\mathbf{w}\bullet\nabla\gamma^s)\,dV$$

$$-\int_{V_\infty}(\boldsymbol{\nu}\bullet\nabla f)\gamma^c(\mathbf{n}\bullet\nabla\gamma)(\mathbf{w}\bullet\nabla\gamma^s)\,dV - \int_{V_\infty} f(\boldsymbol{\nu}\bullet\nabla\gamma^c)(\mathbf{n}\bullet\nabla\gamma)(\mathbf{w}\bullet\nabla\gamma^s)\,dV$$

$$-\int_{V_\infty} f\gamma^c(\boldsymbol{\nu}\bullet\nabla\mathbf{n})\bullet(\nabla\gamma)(\mathbf{w}\bullet\nabla\gamma^s)\,dV - \int_{V_\infty} f\gamma^c\mathbf{n}\bullet[(\nabla\nabla\gamma)\bullet\boldsymbol{\nu}](\mathbf{w}\bullet\nabla\gamma^s)\,dV$$

(6.94)

The first two integrals on the right side may be easily transformed to integrals over a curve using equation (4.11). The third integral is zero because $\boldsymbol{\nu}\bullet\nabla\gamma^c$

is zero by Identity 6.2b (equation (6.11b)). The fourth integral is also zero because, by equation (6.1b), $\nabla\mathbf{n}\bullet\nabla\gamma = 0$. The fifth integral may be simplified if equation (6.27a), $(\nabla\nabla\gamma)\bullet\boldsymbol{\nu} = -\nabla\boldsymbol{\nu}\bullet\nabla\gamma$, is applied and then the resulting expression is transformed to an integral over a curve using equation (4.11). After all these manipulations, equation (6.94) reduces to:

$$\int_{V_\infty} f\gamma^c[\boldsymbol{\nu}\bullet\nabla(\mathbf{w}\bullet\nabla\gamma^s)]\,(\mathbf{n}\bullet\nabla\gamma)\,dV = -\int_C(\nabla\bullet\boldsymbol{\nu})f\mathbf{w}\bullet\boldsymbol{\nu}\,dC - \int_C\mathbf{w}\bullet\boldsymbol{\nu}\boldsymbol{\nu}\bullet\nabla f\,dC$$

$$+\int_C f(\mathbf{n}\bullet\nabla\boldsymbol{\nu}\bullet\mathbf{n})\,\mathbf{w}\bullet\boldsymbol{\nu}\,dC \qquad \textbf{(6.95)}$$

Similar manipulations to those of equations (6.93) through (6.95) applied to the last term in equation (6.92) gives:

$$\int_{V_\infty} f\gamma^c(\boldsymbol{\nu}\bullet\nabla\gamma^s)\,[\mathbf{n}\bullet\nabla(\mathbf{w}\bullet\nabla\gamma)]\,dV = -\int_C(\nabla\bullet\mathbf{n})f\mathbf{w}\bullet\mathbf{n}\,dC - \int_C\mathbf{w}\bullet\mathbf{n}\mathbf{n}\bullet\nabla f\,dC$$

$$+\int_C f(\boldsymbol{\nu}\bullet\nabla\mathbf{n}\bullet\boldsymbol{\nu})\,\mathbf{w}\bullet\mathbf{n}\,dC \qquad \textbf{(6.96)}$$

Substitution of these last two equations back into equation (6.92) and regrouping of terms yields:

$$\int_{V_\infty}\left.\frac{\partial f}{\partial t}\right|_{\mathbf{x}}\gamma^c(\boldsymbol{\nu}\bullet\nabla^s\gamma^s)\,(\mathbf{n}\bullet\nabla\gamma)\,dV = \frac{d}{dt}\int_C f\,dC - f\mathbf{w}\bullet\mathbf{e}\big|_{\text{ends}}$$

$$-\int_C(\nabla\bullet\mathbf{n}-\boldsymbol{\nu}\bullet\nabla\mathbf{n}\bullet\boldsymbol{\nu})f\mathbf{w}\bullet\mathbf{n}\,dC - \int_C(\nabla\bullet\boldsymbol{\nu}-\mathbf{n}\bullet\nabla\boldsymbol{\nu}\bullet\mathbf{n})f\mathbf{w}\bullet\boldsymbol{\nu}\,dC$$

$$-\int_C(\mathbf{w}\bullet\boldsymbol{\nu}\boldsymbol{\nu}+\mathbf{w}\bullet\mathbf{n}\mathbf{n})\bullet\nabla f\,dC \qquad \textbf{(6.97)}$$

Equation (2.38) provides the expression for the divergence of a vector such that:

$$\nabla\bullet\mathbf{n} = \boldsymbol{\lambda}\bullet\nabla\mathbf{n}\bullet\boldsymbol{\lambda}+\mathbf{n}\bullet\nabla\mathbf{n}\bullet\mathbf{n}+\boldsymbol{\nu}\bullet\nabla\mathbf{n}\bullet\boldsymbol{\nu} \qquad \textbf{(6.98a)}$$

$$\nabla\bullet\boldsymbol{\nu} = \boldsymbol{\lambda}\bullet\nabla\boldsymbol{\nu}\bullet\boldsymbol{\lambda}+\mathbf{n}\bullet\nabla\boldsymbol{\nu}\bullet\mathbf{n}+\boldsymbol{\nu}\bullet\nabla\boldsymbol{\nu}\bullet\boldsymbol{\nu} \qquad \textbf{(6.98b)}$$

Recall that from equation (2.12), $\nabla\mathbf{n}\bullet\mathbf{n} = \nabla\boldsymbol{\nu}\bullet\boldsymbol{\nu} = 0$. Furthermore, the velocity of the curve is given by:

$$\mathbf{w} = \mathbf{w}\bullet\boldsymbol{\lambda}\boldsymbol{\lambda} + \mathbf{w}\bullet\boldsymbol{\nu}\boldsymbol{\nu} + \mathbf{w}\bullet\mathbf{n}\mathbf{n} \tag{6.98c}$$

Substitution of these expressions into equation (6.97) yields:

$$\int_{V_\infty} \left.\frac{\partial f}{\partial t}\right|_{\mathbf{x}} \gamma^c (\boldsymbol{\nu}\bullet\nabla^s\gamma^s)\,(\mathbf{n}\bullet\nabla\gamma)\,dV = \frac{d}{dt}\int_C f dC - f\mathbf{w}\bullet\mathbf{e}|_{\text{ends}} - \int_C (\boldsymbol{\lambda}\bullet\nabla\mathbf{n}\bullet\boldsymbol{\lambda}) f\mathbf{w}\bullet\mathbf{n} dC$$

$$- \int_C (\boldsymbol{\lambda}\bullet\nabla\boldsymbol{\nu}\bullet\boldsymbol{\lambda}) f\mathbf{w}\bullet\boldsymbol{\nu} dC - \int_C \mathbf{w}\bullet\nabla f dC + \int_C \mathbf{w}\bullet\boldsymbol{\lambda}\boldsymbol{\lambda}\bullet\nabla f dC \tag{6.99}$$

Equations (2.14b) and (2.14c) indicate, respectively, that $\nabla\mathbf{n}\bullet\boldsymbol{\lambda} = -\nabla\boldsymbol{\lambda}\bullet\mathbf{n}$ and that $\nabla\boldsymbol{\nu}\bullet\boldsymbol{\lambda} = -\nabla\boldsymbol{\lambda}\bullet\boldsymbol{\nu}$. Thus equation (6.99) may be rearranged to the form:

$$\int_{V_\infty} \left.\frac{\partial f}{\partial t}\right|_{\mathbf{x}} \gamma^c (\boldsymbol{\nu}\bullet\nabla^s\gamma^s)\,(\mathbf{n}\bullet\nabla\gamma)\,dV = \frac{d}{dt}\int_C f dC - f\mathbf{w}\bullet\mathbf{e}|_{\text{ends}}$$

$$+ \int_C (\boldsymbol{\lambda}\bullet\nabla\boldsymbol{\lambda})\bullet\mathbf{w} f dC - \int_C \mathbf{w}\bullet\nabla f dC + \int_C \mathbf{w}\bullet\boldsymbol{\lambda}\boldsymbol{\lambda}\bullet\nabla f dC \tag{6.100}$$

where the fact that $\boldsymbol{\lambda}\bullet\nabla\boldsymbol{\lambda}\bullet\,(\mathbf{n}\mathbf{n}\bullet\mathbf{w} + \boldsymbol{\nu}\boldsymbol{\nu}\bullet\mathbf{w}) = \boldsymbol{\lambda}\bullet\nabla\boldsymbol{\lambda}\bullet\mathbf{w}$ has been employed.

Finally, substitution of equation (6.100) back into equation (6.89) and rearrangement of terms yields:

$$\int_C \left.\frac{\partial f}{\partial t}\right|_{\mathbf{u}} dC = \frac{d}{dt}\int_C f dC - \int_C \mathbf{w}\bullet\boldsymbol{\nu}\boldsymbol{\nu}\bullet\nabla f dC + \int_C (\boldsymbol{\lambda}\bullet\nabla\boldsymbol{\lambda})\bullet\mathbf{w} f dC - (\mathbf{e}\bullet\mathbf{w}f)|_{\text{ends}} \tag{6.101}$$

If notation is now employed whereby $\mathbf{w}\bullet\boldsymbol{\nu}\boldsymbol{\nu} \equiv \mathbf{w}^\nu$ and $\boldsymbol{\nu}\boldsymbol{\nu}\bullet\nabla \equiv \nabla^\nu$, then $\mathbf{w}\bullet\boldsymbol{\nu}\boldsymbol{\nu}\bullet\nabla = \mathbf{w}^\nu\bullet\nabla^\nu$ and equation (6.101) may be expressed as:

T[2, (0, 0), 1]

$$\int_C \left.\frac{\partial f}{\partial t}\right|_{\mathbf{u}} dC = \frac{d}{dt}\int_C f dC - \int_C \mathbf{w}^\nu\bullet\nabla^\nu f dC + \int_C (\boldsymbol{\lambda}\bullet\nabla\boldsymbol{\lambda})\bullet\mathbf{w} f dC - (\mathbf{e}\bullet\mathbf{w}f)|_{\text{ends}} \tag{6.102}$$

The derivation in this section is one of the most complex that results when using generalized functions to prove averaging and integration theorems. An alternative derivation that is somewhat simpler would be to develop $T[3,\ (0,0),1]$ and then use equation (6.87) to replace the partial time derivative in space with the partial time derivative on a surface. The result of this procedure, of course, is equation (6.102).

6.6 CONCLUSION

The objectives of this chapter have been to provide a number of identities involving generalized functions and their derivatives and to demonstrate the use of these identities in proving averaging and integration theorems. The relative simplicity of the use of generalized functions in deriving integral theorems was illustrated by the derivation of the well-known Gauss divergence theorem, $D[3,\ (0,0),3]$. The step by step derivation of averaging theorems involving two megascopic and one macroscopic dimensions, the $[3,\ (1,0),2]$ family, further illustrates the use of generalized functions and the manipulations required to obtain the theorems. Finally, a relatively complex derivation was presented that involved the transport theorem for a surface time derivative integrated over a curve in the surface, $T[2,\ (0,0),1]$.

In the next chapter, the full set of integration theorems obtained using the methods of this chapter will be presented without proof. Averaging theorems are given in Chapter 8. Some example applications of the use of these theorems appear in Chapter 9.

6.7 REFERENCES

Batchelor, G. K., *An Introduction to Fluid Mechanics*, Cambridge University Press, Cambridge, 1967.

Eringen, A. C., *Mechanics of Continua*, Krieger, New York, 1980.

Kreyszig, E., *Advanced Engineering Mathematics*, Wiley, New York, 1962.

Slattery, J. C., *Momentum, Energy, and Mass Transfer in Continua*, McGraw-Hill, New York, 1972.

Stone, H. A., A simple derivation of the time-dependent convective-diffusion equation for surfactant transport along a deforming interface, *Physics of Fluids A*, **2**(1), 111-112, 1990.

Whitaker, S., *Introduction to Fluid Mechanics*, Prentice-Hall, Englewood Cliffs, 1968.

Zill, D. G., and M. R. Cullen, *Advanced Engineering Mathematics*, PWS-Kent, Boston, 1992.

CHAPTER SEVEN

INTEGRATION THEOREMS

7.0 INTRODUCTION

The generalized functions described in the previous chapters can be employed to facilitate the derivation of equations useful for a variety of engineering applications. Here attention is focused on presenting theorems which allow for the interchange of the order of integration and differentiation. These theorems are particularly useful, for example, in fluid mechanics problems where it is desirable to change any or all spatial scales from the microscale to the megascale. Some useful identities as well as example derivations have been presented previously in Chapter 6. Based on those principles, the derivation of the remaining theorems may be accomplished. The present chapter will merely present the theorems which have been derived using the generalized functions and explain the notation used, as appropriate. Four kinds of integration theorems will be presented: gradient (G), divergence (D), curl (C), and transport (T) theorems.

The G, D, and C theorems are taken here to include those theorems which relate an integral over a volume, surface, or curve of the gradient, divergence, or curl of a continuous function to integrals of the function over the region and the boundary of the integration region. Thus integrals of microscopic derivatives are converted to megascopic counterparts. The gradient, divergence and curl theorems comprise, respectively, the sets $G[i,\ (0,0), k]$, $D[i,\ (0,0), k]$, and $C[i,\ (0,0), k]$ where $i = 1, 2$, or 3 and $k = 1, 2$, or 3. In practice, the most commonly encountered of these theorems is the set with $i = k = 3$ that transforms the integral over a volume of some spatial derivative of a function to an integral of that function over the bounding surface of the volume. Note that in the defining expression of indices for this class of theorems, $[i,\ (0,0), k]$, the transformation from microscopic coordinates to megascopic coordinates encompasses all the coordinate directions of interest when $i = k$. When $k < i$, microscopic variation in some directions is preserved. When $k > i$, the i-dimensional del operator is integrated over k-space. For example, the theorem $G[3, (0, 0), 1]$ is the form to use when a three-dimensional gradient of a function is to be transformed to a vertically averaged gradient field with explicit variation in the horizontal directions still considered. Theorems presented here from the $[1, (0, 0), 2]$ family are for the case when the single component of the del operator being considered is in a direction tangent to the surface of integration.

Transport theorems relate the integral of a time derivative to the time derivative of the integral. These theorems are characterized by integration of a partial time derivative over a volume, surface, or curve. In other words, integration is over three spatial coordinates, two surficial coordinates, or one curvilineal coordinate, respectively. The integration involves transforming (i.e. integrating) from microscopic coordinates to some mix of microscopic and megascopic spatial coordinates. By the notation introduced, transport theorems are indicated as $T[i, (0, 0), k]$ where $i = 1, 2,$ or 3 and $k = 1, 2,$ or 3. Transport theorems do not involve a macroscopic dimension. For example, after integration over a surface of a partial time derivative evaluated holding all spatial coordinates fixed, the resulting expression will be megascopic in two dimensions but will retain dependence on the microscopic coordinate normal to the surface. The transport theorems are useful tools for single phase fluid mechanics. Typically if a balance equation in "i" spatial dimensions is to be converted to a form involving "k" megascopic dimensions and "$i - k$" microscopic dimensions, the theorem family $[i, (0, 0), k]$ would be employed.

Forms of the theorems are presented first for integration over volumes, then over surfaces, and finally over curves. The forms selected for presentation are those that seem to be particularly convenient for use in analysis of physical problems. Additionally, forms are presented for the special cases when a surface of integration is a plane fixed in space and when a curve of integration is a straight line fixed in space. Volumes of integration are taken to be bounded by closed regular surfaces; surfaces of integration are considered to be orientable; and curves of integration are simple curves. Functions being integrated are assumed to be continuous and to have continuous first partial derivatives.

7.1 THEOREMS FOR INTEGRATION OVER A VOLUME, [*i*, (0, 0), 3]

Typical cases of integration over a volume involve the spatial del operator or the partial time derivative with spatial coordinates held constant in the integrand. These cases correspond to the $[3, (0, 0), 3]$ family and involve full conversion of the microscopic operators to the megascale. Volumetric integration of a differential operator of reduced dimensionality may also be considered. A typical volume of integration, V, with boundary surface, S, is depicted in figure 7.1. The volume is allowed to translate and deform with time.

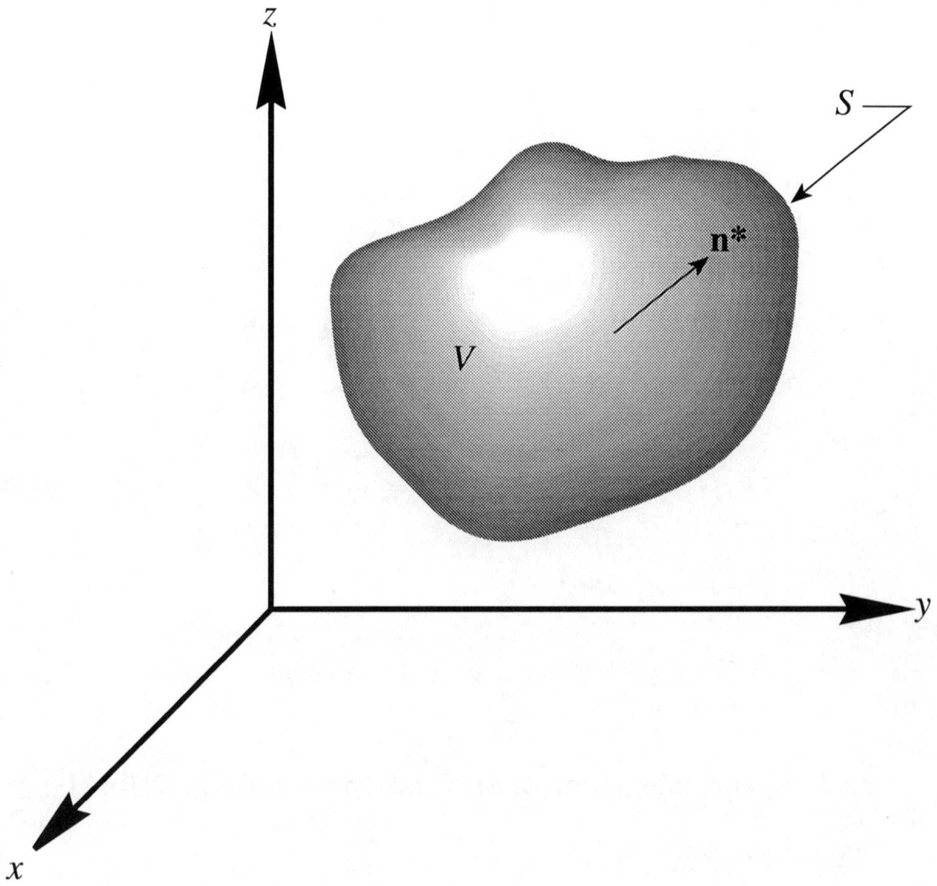

Figure 7.1. Volume of integration, V, with boundary surface, S. The unit vector $\mathbf{n}^*$ is normal to S and directed outward from V.

7.1.1 Spatial Integration of Spatial Operations, [3, (0, 0), 3]

Figure and Notation

Refer to figure 7.1.

V is the volume of integration,

S is the surface boundary of the volume of integration,

$\mathbf{n}^*$ is the unit normal vector outward from V on surface S,

$\mathbf{w}$ is the velocity of the surface,

f is a spatial function $f(\mathbf{x}, t)$, and

$\mathbf{f}$ is a spatial vector $\mathbf{f}(\mathbf{x}, t)$.

G[3, (0, 0), 3] Gradient

$$\int_V \nabla f \, dV = \int_S \mathbf{n}^* f \, dS \qquad (7.1)$$

D[3, (0, 0), 3] Divergence

$$\int_V \nabla \cdot \mathbf{f} \, dV = \int_S \mathbf{n}^* \cdot \mathbf{f} \, dS \qquad (7.2)$$

C[3, (0, 0), 3] Curl

$$\int_V \nabla \times \mathbf{f} \, dV = \int_S \mathbf{n}^* \times \mathbf{f} \, dS \qquad (7.3)$$

T[3, (0, 0), 3] Transport

$$\int_V \left. \frac{\partial f}{\partial t} \right|_{\mathbf{x}} dV = \frac{d}{dt} \int_V f \, dV - \int_S \mathbf{n}^* \cdot \mathbf{w} f \, dS \qquad (7.4)$$

7.1.2 Spatial Integration of Surficial Operations, [2, (0, 0), 3]

Figure and Notation

Refer to figure 7.1.

V is the volume of integration,

S is the surface boundary of the volume of integration,

$\mathbf{n}$ is a unit vector in one of the general orthogonal coordinate directions,

$\mathbf{n}^*$ is the unit normal vector outward from V on surface S,

∇^c is the curvilineal operator such that $\nabla^c = \mathbf{nn} \cdot \nabla$,

∇^s is the surficial operator such that $\nabla^s = \nabla - \nabla^c$,

$\mathbf{w}$ is the velocity of surface S,

$\mathbf{w}^c$ is the velocity of the coordinates orthogonal to $\mathbf{n}$ in direction $\mathbf{n}$ such that $\mathbf{w}^c \times \mathbf{n} = 0$,

f is a spatial function $f(\mathbf{x}, t)$,

$\mathbf{f}$ is a spatial vector $\mathbf{f}(\mathbf{x}, t)$,

$\mathbf{f}^c$ is the vector component of $\mathbf{f}$ tangent to the $\mathbf{n}$-coordinate direction such that $\mathbf{f}^c = \mathbf{nn} \cdot \mathbf{f}$, and

$\mathbf{f}^s$ is the vector component of $\mathbf{f}$ normal to the $\mathbf{n}$-coordinate direction such that $\mathbf{f}^s = \mathbf{f} - \mathbf{f}^c$.

G[2, (0, 0), 3] Gradient

$$\int_V \nabla^s f dV = \int_V (\nabla^s \bullet \mathbf{n})\,\mathbf{n} f dV + \int_V (\mathbf{n} \bullet \nabla^c \mathbf{n}) f dV - \int_S (\mathbf{n}^* \bullet \mathbf{n})\,\mathbf{n} f dS + \int_S \mathbf{n}^* f dS \tag{7.5}$$

D[2, (0, 0), 3] Divergence

$$\int_V \nabla^s \bullet \mathbf{f} dV = \int_V (\nabla^s \bullet \mathbf{n})\,\mathbf{n} \bullet \mathbf{f}^c dV + \int_V (\mathbf{n} \bullet \nabla^c \mathbf{n}) \bullet \mathbf{f}^s dV + \int_S \mathbf{n}^* \bullet \mathbf{f}^s dS \tag{7.6}$$

C[2, (0, 0), 3] Curl

$$\int_V \nabla^s \times \mathbf{f} dV = \int_V (\nabla^s \bullet \mathbf{n})\,\mathbf{n} \times \mathbf{f}^s dV + \int_V (\mathbf{n} \bullet \nabla^c \mathbf{n}) \times \mathbf{f} dV$$

$$- \int_S (\mathbf{n}^* \bullet \mathbf{n})\,\mathbf{n} \times \mathbf{f}^s dS - \int_S \mathbf{n}^* \times \mathbf{f} dS \tag{7.7}$$

T[2, (0, 0), 3] Transport

$$\int_V \left.\frac{\partial f}{\partial t}\right|_{\mathbf{u}} dV = \frac{d}{dt}\int_V f dV - \int_V f \nabla \bullet \mathbf{w}^c dV + \int_S \mathbf{n}^* \bullet \mathbf{w}^c f dS - \int_S \mathbf{n}^* \bullet \mathbf{w} f dS \tag{7.8}$$

7.1.3 Spatial Integration of Curvilineal Operations, [1, (0, 0), 3]

Figure and Notation

Refer to figure 7.1.

- V is the volume of integration,
- S is the surface boundary of the volume of integration,
- $\boldsymbol{\lambda}$ is a unit vector tangent to one of the general orthogonal coordinate directions,
- $\mathbf{n}^*$ is the unit normal vector outward from V on surface S,
- ∇^c is the curvilineal operator such that $\nabla^c = \boldsymbol{\lambda}\boldsymbol{\lambda} \bullet \nabla$,
- ∇^s is the surficial operator such that $\nabla^s = \nabla - \nabla^c$,
- $\mathbf{w}$ is the velocity of surface S,
- $\mathbf{w}^s$ is the velocity of the $\boldsymbol{\lambda}$ coordinate in the direction normal to $\boldsymbol{\lambda}$ such that $\mathbf{w}^s \bullet \boldsymbol{\lambda} = 0$,
- f is a spatial function $f(\mathbf{x}, t)$,

f is a spatial vector $\mathbf{f}(\mathbf{x}, t)$,

$\mathbf{f}^c$ is the vector component of **f** tangent to the $\boldsymbol{\lambda}$-coordinate direction such that $\mathbf{f}^c = \boldsymbol{\lambda}\boldsymbol{\lambda}\cdot\mathbf{f}$, and

$\mathbf{f}^s$ is the vector component of **f** normal to the $\boldsymbol{\lambda}$-coordinate direction such that $\mathbf{f}^s = \mathbf{f} - \mathbf{f}^c$.

G[1, (0, 0), 3] Gradient

$$\int_V \nabla^c f dV = -\int_V (\nabla^s\cdot\boldsymbol{\lambda})\,\boldsymbol{\lambda} f dV - \int_V (\boldsymbol{\lambda}\cdot\nabla^c\boldsymbol{\lambda}) f dV + \int_S (\mathbf{n}^*\cdot\boldsymbol{\lambda})\,\boldsymbol{\lambda} f dS \qquad (7.9)$$

D[1, (0, 0), 3] Divergence

$$\int_V \nabla^c\cdot\mathbf{f} dV = -\int_V (\nabla^s\cdot\boldsymbol{\lambda})\,\boldsymbol{\lambda}\cdot\mathbf{f}^c dV - \int_V (\boldsymbol{\lambda}\cdot\nabla^c\boldsymbol{\lambda})\cdot\mathbf{f}^s dV + \int_S \mathbf{n}^*\cdot\mathbf{f}^c dS \qquad (7.10)$$

C[1, (0, 0), 3] Curl

$$\int_V \nabla^c\times\mathbf{f} dV = -\int_V (\nabla^s\cdot\boldsymbol{\lambda})\,\boldsymbol{\lambda}\times\mathbf{f}^s dV - \int_V (\boldsymbol{\lambda}\cdot\nabla^c\boldsymbol{\lambda})\times\mathbf{f} dV + \int_S (\mathbf{n}^*\cdot\boldsymbol{\lambda})\,\boldsymbol{\lambda}\times\mathbf{f}^s dS \qquad (7.11)$$

T[1, (0, 0), 3] Transport

$$\int_V \left.\frac{\partial f}{\partial t}\right|_1 dV = \frac{d}{dt}\int_V f dV - \int_V f\nabla\cdot\mathbf{w}^s dV + \int_S \mathbf{n}^*\cdot\mathbf{w}^s f dS - \int_S \mathbf{n}^*\cdot\mathbf{w} f dS \qquad (7.12)$$

7.2 THEOREMS FOR INTEGRATION OVER A SURFACE, [*i*, (0, 0), 2]

For integration over a surface, the differential operator considered will be either a spatial operator, a surficial operator in the surface of integration, or a curvilineal operator acting along a curve in the surface. These operators will be considered to be acting on continuous spatial functions with continuous first derivatives. Thus the index "i" in this set of theorems may equal either 3, 2, or 1, though the latter case is likely of somewhat limited interest. A typical surface of integration, S, with boundary curve, C, is depicted in figure 7.2. This surface is allowed to deform and translate with time. Integration over a planar area, A, fixed in space with boundary curve C is a special case of particular utility. Such a surface is presented in figure 7.3. Although the plane is constrained such that it does not move in the direction of its normal, the boundary curve is allowed to move such that the size of the plane is a function of time.

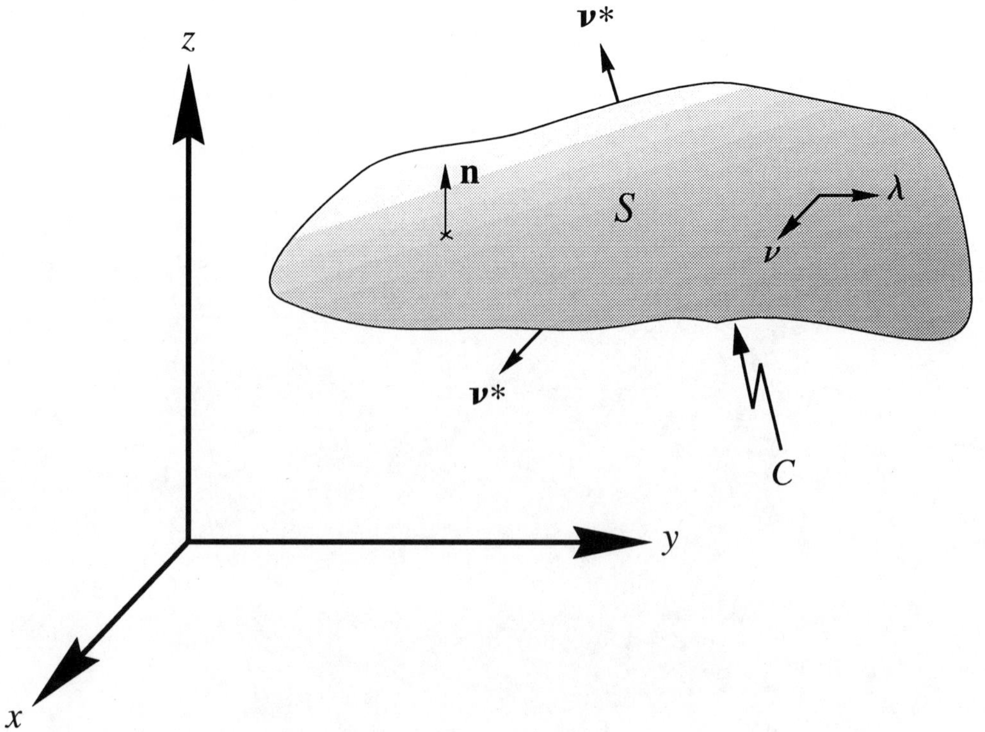

Figure 7.2. Surface of integration, S, with boundary curve, C. The unit vector $\boldsymbol{\nu}^*$ is normal to C, tangent to S, and directed outward from S. The unit vector $\mathbf{n}$ is normal to S.

7.2.1 Surficial Integration of Spatial Operations, [3, (0, 0), 2]

Figure and Notation

Refer to figure 7.2.

S	is the surface of integration,
C	is the curve bounding the surface of integration,
$\mathbf{n}$	is the unit vector normal to surface S,
$\boldsymbol{\nu}^*$	is the unit vector normal to curve C positive in the direction outward from S such that $\mathbf{n}\cdot\boldsymbol{\nu}^* = 0$,
∇^c	is the curvilineal operator such that $\nabla^c = \mathbf{nn}\cdot\nabla$,
∇^s	is the surficial operator such that $\nabla^s = \nabla - \nabla^c$,
$\mathbf{w}$	is the velocity of S,
$\mathbf{w}^c$	is the normal component of $\mathbf{w}$ such that $\mathbf{w}^c = \mathbf{nn}\cdot\mathbf{w}$,

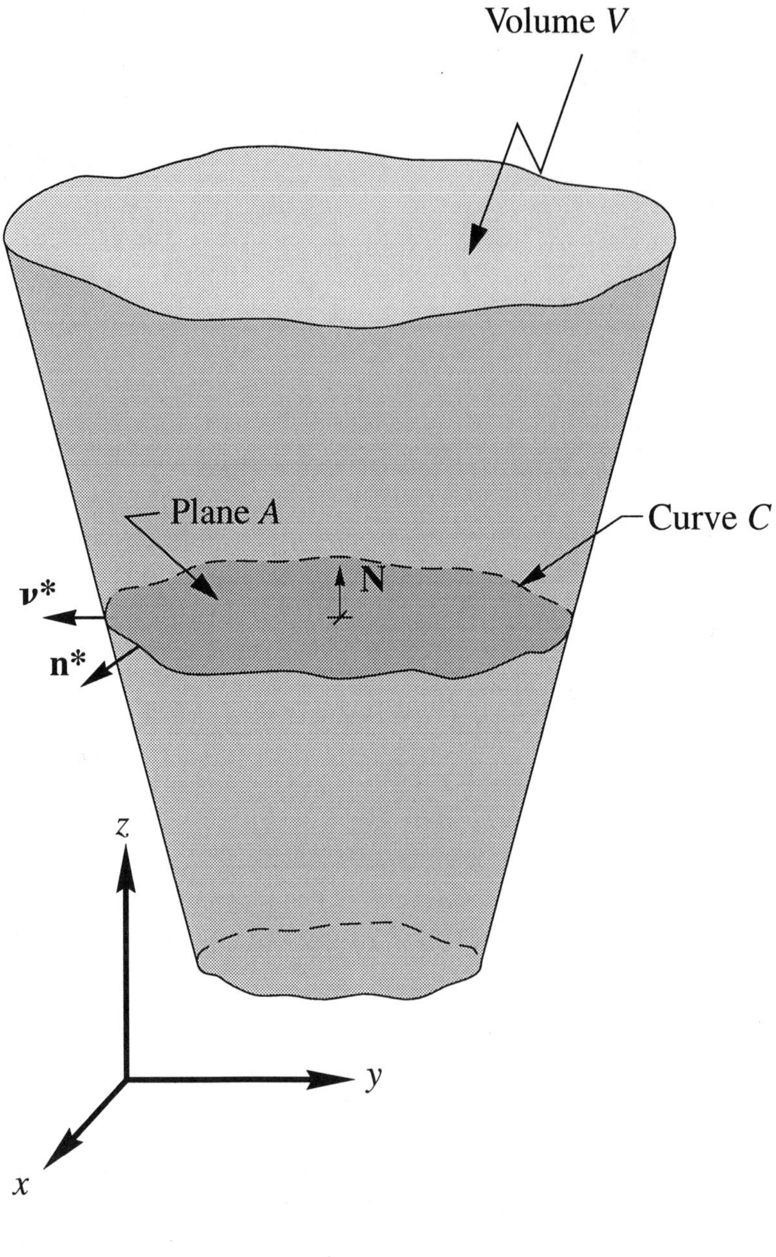

Figure 7.3. Plane of integration, A, with boundary curve, C, that is the intersection of A with the boundary of volume, V. The unit vector $\boldsymbol{\nu}^*$ is normal to C, tangent to A, and directed outward from V. The unit vector $\mathbf{n}^*$ is normal to C and to the boundary of V. The unit vector $\mathbf{N}$ is normal to A.

f is a spatial function $f(\mathbf{x}, t)$,

$\mathbf{f}$ is a spatial vector $\mathbf{f}(\mathbf{x}, t)$,

$\mathbf{f}^c$ is a vector normal to S such that $\mathbf{f}^c = \mathbf{nn}\bullet\mathbf{f}$, and

$\mathbf{f}^s$ is a vector tangent to S such that $\mathbf{f}^s = \mathbf{f} - \mathbf{f}^c$.

G[3, (0, 0), 2] Gradient

$$\int_S \nabla f dS = \int_S \nabla^c f dS + \int_S (\nabla^s \bullet \mathbf{n})\, \mathbf{n} f dS + \int_C \boldsymbol{\nu}^* f dC \tag{7.13}$$

D[3, (0, 0), 2] Divergence

$$\int_S \nabla\bullet\mathbf{f} dS = \int_S \nabla^c \bullet \mathbf{f}^c dS - \int_S \mathbf{n}\bullet\nabla^c \mathbf{n}\bullet\mathbf{f}^s dS + \int_S (\nabla^s\bullet\mathbf{n})\, \mathbf{n}\bullet\mathbf{f}^c dS + \int_C \boldsymbol{\nu}^* \bullet \mathbf{f} dC \tag{7.14}$$

C[3, (0, 0), 2] Curl

$$\int_S \nabla\times\mathbf{f} dS = \int_S \nabla^c \times \mathbf{f}^s dS - \int_S (\mathbf{n}\bullet\nabla^c \mathbf{n}) \times \mathbf{f}^c dS + \int_S (\nabla^s\bullet\mathbf{n})\, \mathbf{n}\times\mathbf{f}^s dS + \int_C \boldsymbol{\nu}^* \times \mathbf{f} dC \tag{7.15}$$

T[3, (0, 0), 2] Transport

$$\int_S \left.\frac{\partial f}{\partial t}\right|_{\mathbf{x}} dS = \frac{d}{dt}\int_S f dS - \int_S (\nabla^s\bullet\mathbf{n})\, \mathbf{w}^c \bullet \mathbf{n} f dS - \int_S \mathbf{w}^c \bullet \nabla^c f dS - \int_C \boldsymbol{\nu}^* \bullet \mathbf{w} f dC \tag{7.16}$$

7.2.2 Integration of Spatial Operations over a Plane Fixed in Space, [3, (0, 0), 2]

Figure and Notation

Refer to figure 7.3.

A is the planar surface of integration fixed in space,

C is the curve bounding the plane of integration and is also the curve of intersection of an extension of plane A with a volume,

$\mathbf{N}$ is the unit vector normal to planar surface A,

$\boldsymbol{\nu}^*$ is the unit vector normal to curve C positive outward from A such that $\mathbf{N}\bullet\boldsymbol{\nu}^* = 0$,

n*   is the unit vector normal to curve C that is also normal to the boundary of a volume intersected by an extension of plane A to form curve C,

∇^c is the curvilineal operator such that $\nabla^c = \mathbf{NN}\bullet\nabla$,

w   is the velocity of C such that $\mathbf{w} = \boldsymbol{\nu}^*\boldsymbol{\nu}^*\bullet\mathbf{w}$,

f is a spatial function $f(\mathbf{x}, t)$,

f   is a spatial vector $\mathbf{f}(\mathbf{x}, t)$,

$\mathbf{f}^c$ is a vector normal to A such that $\mathbf{f}^c = \mathbf{NN}\bullet\mathbf{f}$, and

$\mathbf{f}^s$ is a vector tangent to A such that $\mathbf{f}^s = \mathbf{f} - \mathbf{f}^c$.

G[3, (0, 0), 2] Gradient

$$\int_A \nabla f dA = \nabla^c \int_A f dA + \int_C \frac{\mathbf{n}^* f}{(\boldsymbol{\nu}^*\bullet\mathbf{n}^*)} dC \quad \textbf{(7.17)}$$

D[3, (0, 0), 2] Divergence

$$\int_A \nabla\bullet\mathbf{f} dA = \nabla^c\bullet\int_A \mathbf{f}^c dA - \int_A (\mathbf{N}\bullet\nabla^c\mathbf{N})\bullet\mathbf{f}^s dA + \int_C \frac{\mathbf{n}^*\bullet\mathbf{f}}{(\boldsymbol{\nu}^*\bullet\mathbf{n}^*)} dC \quad \textbf{(7.18)}$$

C[3, (0, 0), 2] Curl

$$\int_A \nabla\times\mathbf{f} dA = \nabla^c\times\int_A \mathbf{f}^s dA - \int_A (\mathbf{N}\bullet\nabla^c\mathbf{N})\times\mathbf{f}^c dA + \int_C \frac{\mathbf{n}^*\times\mathbf{f}}{(\boldsymbol{\nu}^*\bullet\mathbf{n}^*)} dC \quad \textbf{(7.19)}$$

T[3, (0, 0), 2] Transport

$$\int_A \left.\frac{\partial f}{\partial t}\right|_{\mathbf{x}} dA = \frac{\partial}{\partial t}\int_A f dA - \int_C \frac{\mathbf{n}^*\bullet\mathbf{w} f}{(\boldsymbol{\nu}^*\bullet\mathbf{n}^*)} dC \quad \textbf{(7.20)}$$

7.2.3 Surficial Integration of Surface Operations, [2, (0, 0), 2]

Figure and Notation

Refer to figure 7.2.

S is the surface of integration,

C is the curve bounding the surface of integration,

n   is the unit vector normal to surface S,

$\boldsymbol{\nu}^*$ is the unit vector normal to curve C positive in the direction outward from S such that $\mathbf{n}\bullet\boldsymbol{\nu}^* = 0$,

∇^s is the surficial operator such that $\nabla^s = \nabla - \mathbf{nn}\bullet\nabla$,

$\mathbf{w}$ is the velocity of S,

f is a surficial function $f(\mathbf{u}, t)$

$\mathbf{f}$ is a surficial vector $\mathbf{f}(\mathbf{u}, t)$,

$\mathbf{f}^c$ is a vector normal to S such that $\mathbf{f}^c = \mathbf{nn}\bullet\mathbf{f}$, and

$\mathbf{f}^s$ is a vector tangent to S such that $\mathbf{f}^s = \mathbf{f} - \mathbf{f}^c$.

G[2, (0, 0), 2] Gradient

$$\int_S \nabla^s f dS = \int_S (\nabla^s\bullet\mathbf{n})\,\mathbf{n} f dS + \int_C \boldsymbol{\nu}^* f dC \tag{7.21}$$

D[2, (0, 0), 2] Divergence

$$\int_S \nabla^s\bullet\mathbf{f} dS = \int_S (\nabla^s\bullet\mathbf{n})\,\mathbf{n}\bullet\mathbf{f}^c dS + \int_C \boldsymbol{\nu}^*\bullet\mathbf{f}^s dC \tag{7.22}$$

C[2, (0, 0), 2] Curl

$$\int_S \nabla^s \times \mathbf{f} dS = \int_S (\nabla^s\bullet\mathbf{n})\,\mathbf{n}\times\mathbf{f}^s dS + \int_C \boldsymbol{\nu}^* \times \mathbf{f} dC \tag{7.23}$$

T[2, (0, 0), 2] Transport

$$\int_S \left.\frac{\partial f}{\partial t}\right|_{\mathbf{u}} dS = \frac{d}{dt}\int_S f dS - \int_S (\nabla^s\bullet\mathbf{n})\,\mathbf{n}\bullet\mathbf{w} f dS - \int_C \boldsymbol{\nu}^*\bullet\mathbf{w} f dC \tag{7.24}$$

7.2.4 Integration of Surficial Operations over a Plane Fixed in Space, [2, (0, 0), 2]

Figure and Notation

Refer to figure 7.3.

A is the planar surface of integration fixed in space,

C is the curve bounding the plane of integration and is also the curve of intersection of an extension of plane A with a volume,

$\mathbf{N}$ is the unit vector normal to planar surface A,

$\boldsymbol{\nu}^*$ is the unit vector normal to curve C positive outward from A such that $\mathbf{N}\bullet\boldsymbol{\nu}^* = 0$,

∇^s is the surficial operator such that $\nabla^s = \nabla - \mathbf{NN}\bullet\nabla$,

w is the velocity of C such that $\mathbf{w} = \boldsymbol{\nu}^* \boldsymbol{\nu}^* \cdot \mathbf{w}$,

f is a surficial function $f(\mathbf{u}, t)$

f is a surficial vector $\mathbf{f}(\mathbf{u}, t)$,

$\mathbf{f}^C$ is a vector normal to A such that $\mathbf{f}^C = \mathbf{N}\mathbf{N} \cdot \mathbf{f}$, and

$\mathbf{f}^S$ is the vector tangent to A such that $\mathbf{f}^S = \mathbf{f} - \mathbf{f}^C$.

G[2, (0, 0), 2] Gradient

$$\int_A \nabla^s f dA = \int_C \boldsymbol{\nu}^* f dC \tag{7.25}$$

D[2, (0, 0), 2] Divergence

$$\int_A \nabla^S \cdot \mathbf{f} dA = \int_C \boldsymbol{\nu}^* \cdot \mathbf{f}^s dC \tag{7.26}$$

C[2, (0, 0), 2] Curl

$$\int_A \nabla^S \times \mathbf{f} dA = \int_C \boldsymbol{\nu}^* \times \mathbf{f} dC \tag{7.27}$$

T[2, (0, 0), 2] Transport

$$\int_A \left.\frac{\partial f}{\partial t}\right|_{\mathbf{u}} dA = \frac{d}{dt}\int_A f dA - \int_C \boldsymbol{\nu}^* \cdot \mathbf{w} f dC \tag{7.28}$$

7.2.5 Surficial Integration of Curvilineal Operations, [1, (0, 0), 2]

Figure and Notation

Refer to figure 7.2.

S is the surface of integration,

C is the curve bounding the surface of integration,

$\boldsymbol{\lambda}$ is the unit vector tangent to one of the surficial coordinates in S,

n is the unit vector normal to surface S,

$\boldsymbol{\nu}^*$ is the unit vector normal to curve C positive in the direction outward from S such that $\mathbf{n} \cdot \boldsymbol{\nu}^* = 0$,

∇^S is the surficial operator such that $\nabla^S = \nabla - \mathbf{n}\mathbf{n} \cdot \nabla$,

∇^λ is the curvilineal operator such that $\nabla^\lambda = \boldsymbol{\lambda}\boldsymbol{\lambda} \cdot \nabla$

w is the velocity of C,

$\mathbf{w}^{\nu}$ is the velocity of the $\boldsymbol{\lambda}$-coordinate in the coordinate direction normal to $\boldsymbol{\lambda}$ such that $\mathbf{w}^{\nu} = \boldsymbol{\nu}\boldsymbol{\nu}\bullet\mathbf{w}^{\nu}$,

f is a surficial function $f(\mathbf{u}, t)$, and

$\mathbf{f}$ is a surficial vector $\mathbf{f}(\mathbf{u}, t)$.

G[1, (0, 0), 2] Gradient

$$\int_S \nabla^{\lambda} f dS = -\int_S (\nabla^s \bullet \boldsymbol{\lambda})\boldsymbol{\lambda} f dS - \int_S (\boldsymbol{\lambda}\bullet\nabla^{\lambda}\boldsymbol{\lambda}) f dS + \int_C \boldsymbol{\nu}^* \bullet \boldsymbol{\lambda}\boldsymbol{\lambda} f dC \quad \textbf{(7.29)}$$

D[1, (0, 0), 2] Divergence

$$\int_S \nabla^{\lambda} \bullet \mathbf{f} dS = -\int_S (\nabla^s \bullet \boldsymbol{\lambda})\boldsymbol{\lambda}\bullet\mathbf{f} dS - \int_S (\boldsymbol{\lambda}\bullet\nabla^{\lambda}\boldsymbol{\lambda})\bullet\mathbf{f} dS + \int_C (\boldsymbol{\nu}^* \bullet \boldsymbol{\lambda})\boldsymbol{\lambda}\bullet\mathbf{f} dC \quad \textbf{(7.30)}$$

C[1, (0, 0), 2] Curl

$$\int_S \nabla^{\lambda} \times \mathbf{f} dS = -\int_S (\nabla^s \bullet \boldsymbol{\lambda})\boldsymbol{\lambda}\times\mathbf{f} dS - \int_S (\boldsymbol{\lambda}\bullet\nabla^{\lambda}\boldsymbol{\lambda})\times\mathbf{f} dS + \int_C (\boldsymbol{\nu}^* \bullet \boldsymbol{\lambda})\boldsymbol{\lambda}\times\mathbf{f} dC \quad \textbf{(7.31)}$$

T[1, (0, 0), 2] Transport

$$\int_S \left.\frac{\partial f}{\partial t}\right|_1 dS = \frac{d}{dt}\int_S f dS - \int_S (\nabla^s \bullet \mathbf{n})\mathbf{n}\bullet\mathbf{w} dS - \int_S f(\nabla^s \bullet \mathbf{w}^{\nu}) dS - \int_C (\boldsymbol{\nu}^* \bullet \boldsymbol{\lambda})\boldsymbol{\lambda}\bullet\mathbf{w} f dC \quad \textbf{(7.32)}$$

7.3 THEOREMS FOR INTEGRATION OVER A CURVE, [*i*, (0, 0), 1]

For integration over a curve, the differential operator in the integrand may be either spatial, surficial, or curvilineal. Thus the index "i" in this set of theorems may equal either 3, 2, or 1. A typical simple curve of integration, C, with two endpoints is depicted in figure 7.4. In figure 7.5, a curve C is presented that lies in a surface. Although the curve in figure 7.4 can also be considered as residing in a surface, the explicit case depicted in figure 7.5 will be notationally convenient. Each of the curves in figure 7.4 and 7.5 is allowed to deform and translate with time. Integration over a straight line segment, L, fixed in space is a special curve of particular utility. Such a line segment is presented in figure 7.6 where its ends are the points of intersection of the straight line with the surface of a

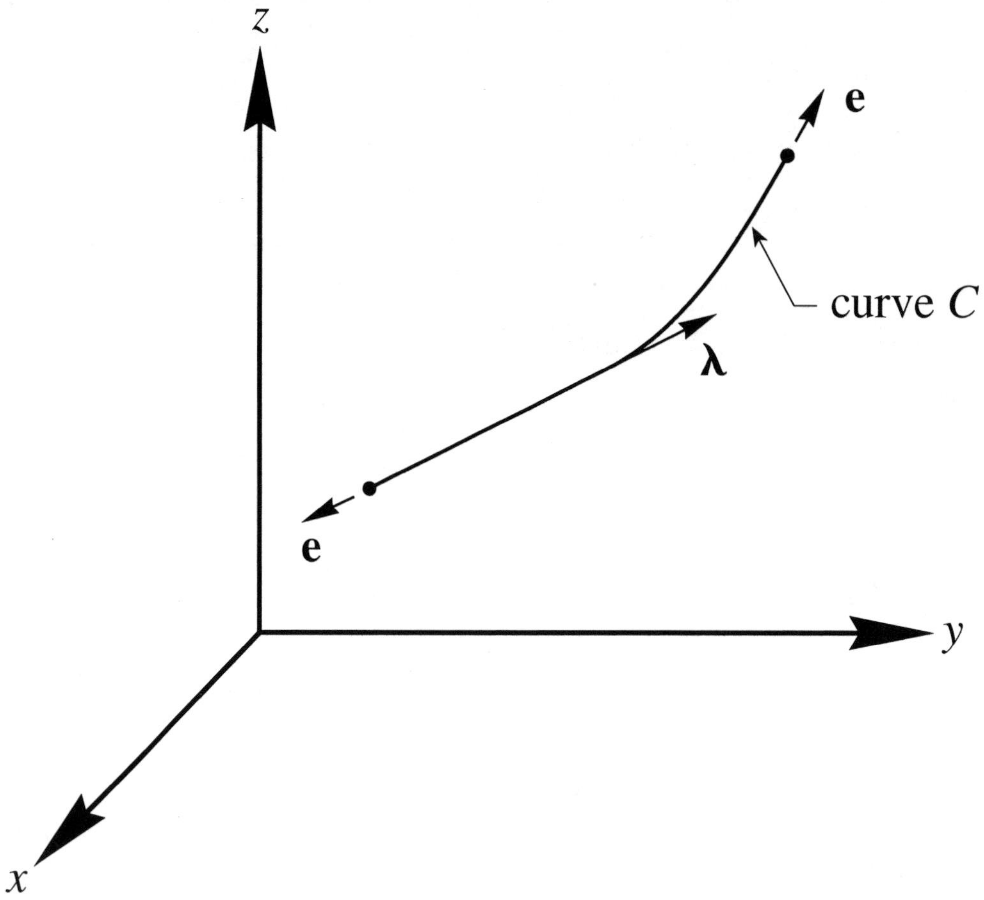

Figure 7.4. Simple curve of integration, C. The unit vector $\boldsymbol{\lambda}$ is tangent to the curve. The unit vectors tangent to the curve and pointing outward from the curve at the endpoints are indicated as $\mathbf{e}$.

volume. A straight line segment lying in a plane is shown in figure 7.7. The ends of this segment are the points of intersection of the line with the curve bounding a plane. Although the line segments in figures 7.6 and 7.7 are constrained such that they do not translate or rotate, the endpoints of the segments may move in directions tangent to the segments such that their lengths are functions of time.

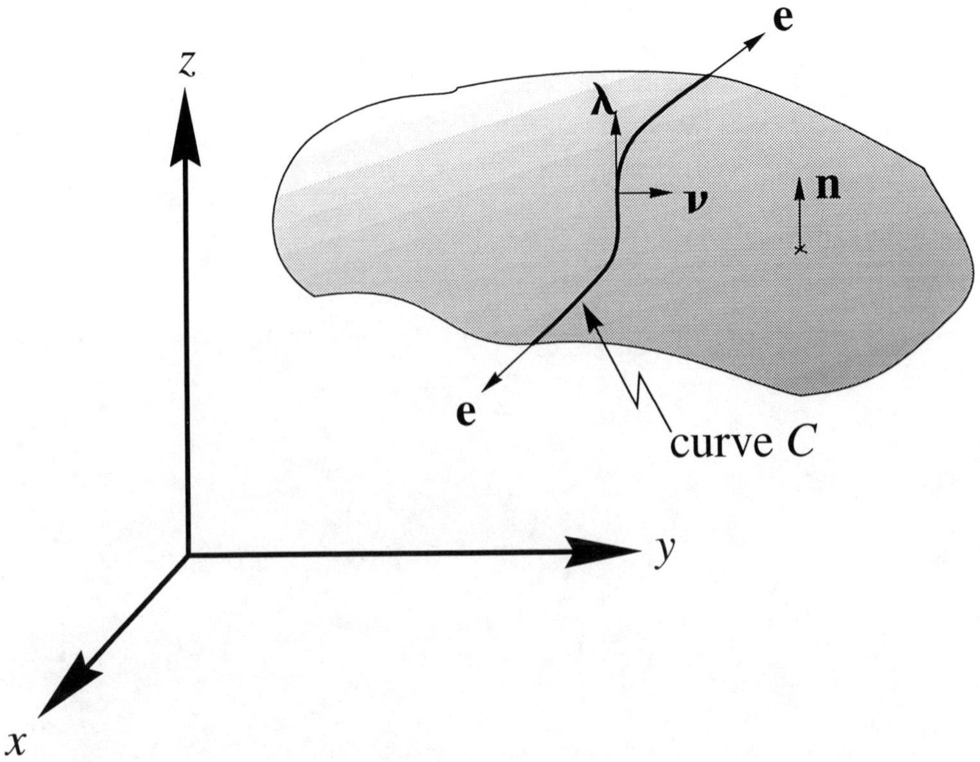

Figure 7.5. Simple curve of integration, C, that lies in a surface. The unit vector $\boldsymbol{\lambda}$ is tangent to the curve; the unit vector $\boldsymbol{\nu}$ is normal to the curve but tangent to the surface; and the unit vector $\mathbf{n}$ is normal to the surface. The unit vectors tangent to the curve and pointing outward from the curve at the endpoints are indicated as $\mathbf{e}$.

7.3.1 Curvilineal Integration of Spatial Operations, [3, (0, 0), 1]

Figure and Notation

Refer to figure 7.4.

C is the curve of integration,

$\boldsymbol{\lambda}$ is the unit vector tangent to curve C,

$\mathbf{e}$ is the unit vector tangent to curve C at the endpoints positive in the direction out from the curve,

∇^c is the curvilineal operator such that $\nabla^c = \boldsymbol{\lambda\lambda}\cdot\nabla$,

∇^s is the surficial operator such that $\nabla^s = \nabla - \nabla^c$,

$\mathbf{w}$ is the velocity of C,

$\mathbf{w}^c$ is the velocity of C tangent to C such that $\mathbf{w}^c = \boldsymbol{\lambda\lambda}\cdot\mathbf{w}$,

- $\mathbf{w}^s$ is the velocity of C normal C such that $\mathbf{w}^s = \mathbf{w} - \mathbf{w}^c$,
- f is a spatial function $f(\mathbf{x}, t)$,
- $\mathbf{f}$ is a spatial vector $\mathbf{f}(\mathbf{x}, t)$,
- $\mathbf{f}^c$ is the vector tangent to C such that $\mathbf{f}^c = \boldsymbol{\lambda}\boldsymbol{\lambda}\cdot\mathbf{f}$, and
- $\mathbf{f}^s$ is the vector normal to C such that $\mathbf{f}^s = \mathbf{f} - \mathbf{f}^c$.

G[3, (0, 0), 1] Gradient

$$\int_C \nabla f dC = \int_C \nabla^s f dC - \int_C (\boldsymbol{\lambda}\cdot\nabla^c\boldsymbol{\lambda}) f dC + \mathbf{e}f|_{\text{ends}} \tag{7.33}$$

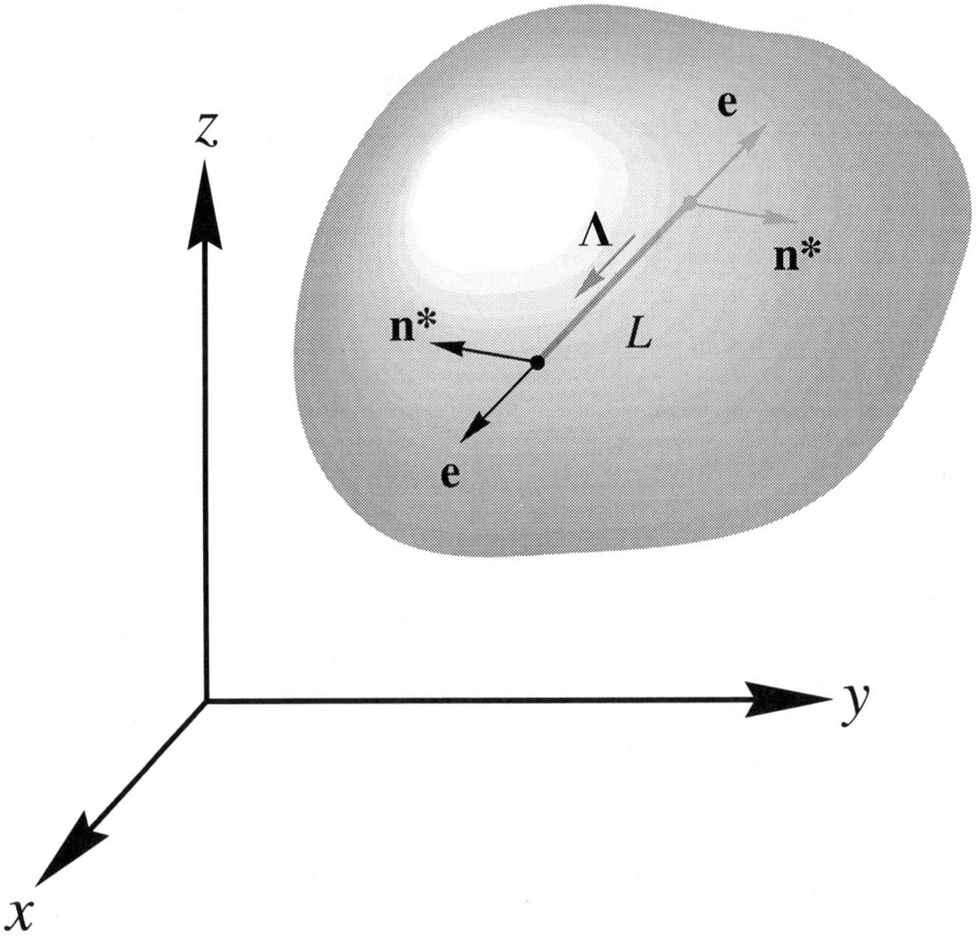

Figure 7.6. A straight line segment, L, whose endpoints are the points of intersection of the line with the surface of a volume. The unit vector $\mathbf{n}^*$ is the outward normal to the surface; the unit vectors tangent to the line and pointing outward from the segment at the endpoints are indicated as $\mathbf{e}$. The unit vector tangent to L is $\boldsymbol{\Lambda}$.

D[3, (0, 0), 1] Divergence

$$\int_C \nabla\cdot\mathbf{f}dC = \int_C \nabla^s\cdot\mathbf{f}^s dC + \int_C (\nabla^s\cdot\boldsymbol{\lambda})\,\boldsymbol{\lambda}\cdot\mathbf{f}^c dC - \int_C (\boldsymbol{\lambda}\cdot\nabla^c\boldsymbol{\lambda})\cdot\mathbf{f}^s dC + (\mathbf{e}\cdot\mathbf{f}^c)\big|_{\text{ends}} \tag{7.34}$$

C[3, (0, 0), 1] Curl

$$\int_C \nabla\times\mathbf{f}dC = \int_C \nabla^s\times\mathbf{f}dC - \int_C (\boldsymbol{\lambda}\cdot\nabla^c\boldsymbol{\lambda})\times\mathbf{f}dC + (\mathbf{e}\times\mathbf{f}^s)\big|_{\text{ends}} \tag{7.35}$$

T[3, (0, 0), 1] Transport

$$\int_C \frac{\partial f}{\partial t}\bigg|_{\mathbf{x}} dC = \frac{d}{dt}\int_C f dC - \int_C \mathbf{w}^s\cdot\nabla^s f dC + \int_C (\boldsymbol{\lambda}\cdot\nabla^c\boldsymbol{\lambda})\cdot\mathbf{w}^s f dC - (\mathbf{e}\cdot\mathbf{w}^c f)\big|_{\text{ends}} \tag{7.36}$$

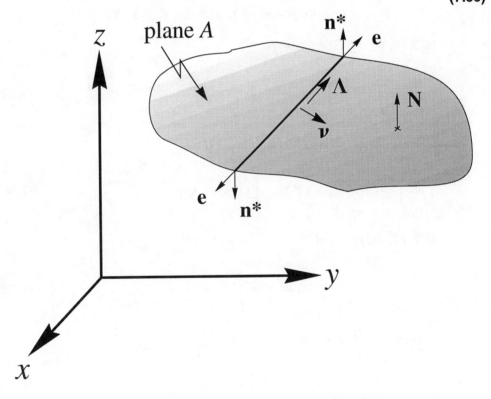

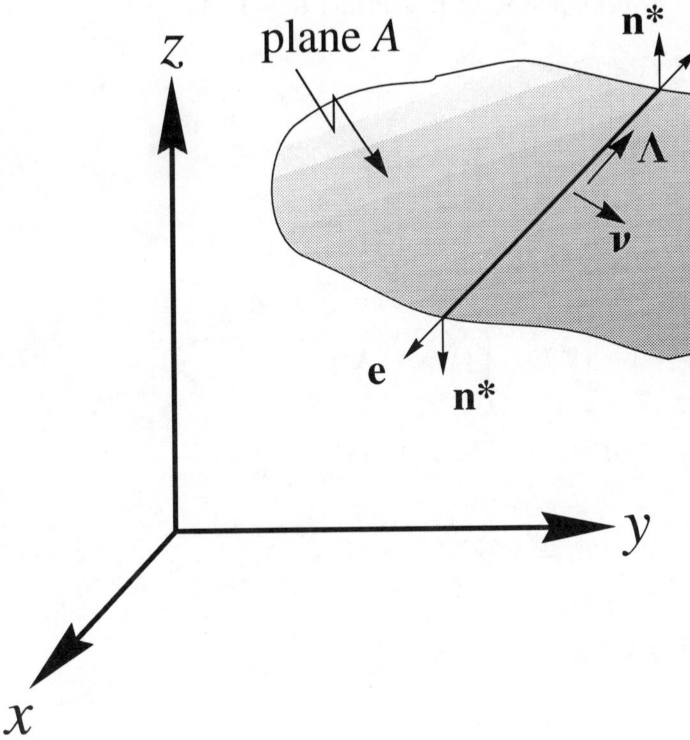

Figure 7.7. Straight line segment, L, contained in plane A. The unit vector $\boldsymbol{\Lambda}$ is tangent to L; unit vector $\boldsymbol{\nu}$ is normal to L and tangent to A; unit vector $\mathbf{N}$ is normal to A; unit vector $\mathbf{n^*}$ is tangent to A and normal to the curve that is the boundary of A; the unit vectors tangent to L and pointing outward from the segment at the endpoints are indicated as $\mathbf{e}$.

7.3.2 Integration of Spatial Operations over a Straight Line Fixed in Space, [3, (0, 0), 1]

Figure and Notation

Refer to figure 7.6.

- L is the straight line segment of integration,
- $\boldsymbol{\Lambda}$ is the unit vector tangent to segment L,
- $\mathbf{e}$ is the unit vector tangent to segment L at its endpoints, positive in the direction out from the line,
- $\mathbf{n}^*$ is the unit vector at the ends of the segment L that is normal to the surface of a volume that is pierced by the straight line,
- ∇^s is the surficial operator such that $\nabla^s = \nabla - \boldsymbol{\Lambda\Lambda}\bullet\nabla$,
- $\mathbf{w}$ is the velocity of endpoints of the line L such that $\mathbf{w} = \boldsymbol{\Lambda\Lambda}\bullet\mathbf{w}$,
- f is a spatial function $f(\mathbf{x}, t)$,
- $\mathbf{f}$ is a spatial vector $\mathbf{f}(\mathbf{x}, t)$,
- $\mathbf{f}^c$ is a vector tangent to L such that $\mathbf{f}^c = \boldsymbol{\Lambda\Lambda}\bullet\mathbf{f}$, and
- $\mathbf{f}^s$ is a vector normal to L such that $\mathbf{f}^s = \mathbf{f} - \mathbf{f}^c$.

G[3, (0, 0), 1] Gradient

$$\int_L \nabla f dL = \nabla^s \int_L f dL + \left(\frac{\mathbf{n}^* f}{\mathbf{e}\bullet\mathbf{n}^*}\right)\Bigg|_{\text{ends}} \tag{7.37}$$

D[3, (0, 0), 1] Divergence

$$\int_L \nabla\bullet\mathbf{f} dL = \nabla^s\bullet\int_L \mathbf{f}^s dL + \int_L (\nabla^s\bullet\boldsymbol{\Lambda})\,\boldsymbol{\Lambda}\bullet\mathbf{f}^c dL + \left(\frac{\mathbf{n}^*\bullet\mathbf{f}}{\mathbf{e}\bullet\mathbf{n}^*}\right)\Bigg|_{\text{ends}} \tag{7.38}$$

C[3, (0, 0), 1] Curl

$$\int_L \nabla\times\mathbf{f} dL = \nabla^s\times\int_L \mathbf{f} dL + \left(\frac{\mathbf{n}^*\times\mathbf{f}}{\mathbf{e}\bullet\mathbf{n}^*}\right)\Bigg|_{\text{ends}} \tag{7.39}$$

T[3, (0, 0), 1] Transport

$$\int_L \frac{\partial f}{\partial t}\bigg|_{\mathbf{x}} dL = \frac{\partial}{\partial t}\int_L f dL - \left(\frac{\mathbf{n}^*\bullet\mathbf{w}f}{\mathbf{e}\bullet\mathbf{n}^*}\right)\Bigg|_{\text{ends}} \tag{7.40}$$

7.3.3 Curvilineal Integration of Surface Operations, [2, (0, 0), 1]

Figure and Notation

Refer to figure 7.5.

C is the curve of integration,

$\boldsymbol{\lambda}$ is the unit vector tangent to curve C,

$\mathbf{e}$ is the unit vector tangent to curve C at the endpoints positive in the direction out from the curve,

$\mathbf{n}$ is the unit vector normal to the surface containing curve C,

$\boldsymbol{\nu}$ is the unit vector normal to curve C and tangent to the surface containing C,

∇^s is the surficial operator such that $\nabla^s = \nabla - \mathbf{nn}\cdot\nabla$,

∇^ν is the curvilineal operator along the curve normal to C and in the surface containing C such that $\nabla^\nu = \nabla^s - \boldsymbol{\lambda\lambda}\cdot\nabla^s$,

$\mathbf{w}$ is the velocity of C,

$\mathbf{w}^s$ is the velocity of C in the surface containing C such that $\mathbf{w}^s = \mathbf{w} - \mathbf{nn}\cdot\mathbf{w}$,

$\mathbf{w}^\nu$ is the velocity of C in the $\boldsymbol{\nu}$ direction such that $\mathbf{w}^\nu = \boldsymbol{\nu\nu}\cdot\mathbf{w}$,

$\mathbf{w}^\lambda$ is the velocity of C in the $\boldsymbol{\lambda}$ direction such that $\mathbf{w}^\lambda = \boldsymbol{\lambda\lambda}\cdot\mathbf{w}$,

f is a spatial function $f(\mathbf{x}, t)$,

$\mathbf{f}$ is a spatial vector $\mathbf{f}(\mathbf{x}, t)$,

$\mathbf{f}^n$ is a vector normal to C and to the surface containing C such that $\mathbf{f}^n = \mathbf{nn}\cdot\mathbf{f}$,

$\mathbf{f}^s$ is a surficial vector such that $\mathbf{f}^s = \mathbf{f} - \mathbf{f}^n$,

$\mathbf{f}^\lambda$ is a vector tangent to C such that $\mathbf{f}^\lambda = \boldsymbol{\lambda\lambda}\cdot\mathbf{f}$, and

$\mathbf{f}^\nu$ is a vector normal to C such that $\mathbf{f}^\nu = \boldsymbol{\nu\nu}\cdot\mathbf{f}$.

G[2, (0, 0), 1] Gradient

$$\int_C \nabla^s f dC = \int_C \nabla^\nu f dC - \int_C \boldsymbol{\lambda}\cdot\nabla\boldsymbol{\lambda} f dC + (\mathbf{e}f)\big|_{\text{ends}} \tag{7.41}$$

D[2, (0, 0), 1] Divergence

$$\int_C \nabla^s\cdot\mathbf{f} dC = \int_C \nabla^\nu\cdot\mathbf{f}^\nu dC - \int_C (\boldsymbol{\lambda}\cdot\nabla\boldsymbol{\lambda})\cdot\mathbf{f} dC - \int_C (\boldsymbol{\nu}\cdot\nabla\boldsymbol{\nu})\cdot\mathbf{f} dC + (\mathbf{e}\cdot\mathbf{f}^\lambda)\big|_{\text{ends}} \tag{7.42}$$

C[2, (0, 0), 1] Curl

$$\int_C \nabla^s \times \mathbf{f} dC = \int_C \nabla^\nu \times (\mathbf{f}^\lambda + \mathbf{f}^n)\, dC - \int_C (\boldsymbol{\nu} \cdot \nabla \boldsymbol{\nu}) \times \mathbf{f}^\nu dC$$

$$- \int_C (\boldsymbol{\lambda} \cdot \nabla \boldsymbol{\lambda}) \times \mathbf{f} dC + [\mathbf{e} \times (\mathbf{f}^n + \mathbf{f}^\nu)]\big|_{\text{ends}} \qquad \textbf{(7.43)}$$

T[2, (0, 0), 1] Transport

$$\int_C \frac{\partial f}{\partial t}\bigg|_{\mathbf{u}} dC = \frac{d}{dt}\int_C f dC - \int_C \mathbf{w}^\nu \cdot \nabla^\nu f dC + \int_C (\boldsymbol{\lambda} \cdot \nabla \boldsymbol{\lambda}) \cdot \mathbf{w} f dC - (\mathbf{e} \cdot \mathbf{w}^\lambda f)\big|_{\text{ends}}$$

(7.44)

7.3.4 Integration of Surface Operations over a Straight Line Fixed on a Planar Surface, [2, (0, 0), 1]

Figure and Notation

Refer to figure 7.7.

- L is the straight line segment of integration,
- $\boldsymbol{\Lambda}$ is the unit vector tangent to segment L,
- $\mathbf{e}$ is the unit vector tangent to segment L at the endpoints positive in the direction out from the segment,
- $\mathbf{N}$ is the unit vector normal to the plane containing segment L,
- $\mathbf{n}^*$ is the unit vector at the intersection of the ends of the segment L with the curve bounding the plane containing L that is normal to the curve and tangent to the plane,
- ∇^s is the surficial operator such that $\nabla^s = \nabla - \mathbf{NN} \cdot \nabla$,
- ∇^ν is the curvilineal operator along the curve normal to L and in the surface containing L such that $\nabla^\nu = \nabla^s - \boldsymbol{\Lambda\Lambda} \cdot \nabla^s$,
- $\mathbf{w}$ is the velocity of the endpoints of L such that $\mathbf{w} = \boldsymbol{\Lambda\Lambda} \cdot \mathbf{w}$,
- f is a surficial function $f(\mathbf{u}, t)$,
- $\mathbf{f}$ is a surficial vector $\mathbf{f}^s(\mathbf{u}, t)$,
- $\mathbf{f}^\lambda$ is the vector tangent to L such that $\mathbf{f}^\lambda = \boldsymbol{\Lambda\Lambda} \cdot \mathbf{f}$,
- $\mathbf{f}^n$ is the vector normal to L and normal to the plane containing L such that $\mathbf{f}^n = \mathbf{NN} \cdot \mathbf{f}$, and
- $\mathbf{f}^\nu$ is the vector normal to L and tangent to the plane containing L such that $\mathbf{f}^\nu = \boldsymbol{\nu\nu} \cdot \mathbf{f}$.

G[2, (0, 0), 1] Gradient

$$\int_L \nabla^s f dL = \nabla^\nu \int_L f dL + \left(\frac{\mathbf{n}^* f}{\mathbf{e} \bullet \mathbf{n}^*}\right)\Bigg|_{\text{ends}} \tag{7.45}$$

D[2, (0, 0), 1] Divergence

$$\int_L \nabla^s \bullet \mathbf{f} dL = \nabla^\nu \bullet \int_L \mathbf{f}^\nu dL + \int_L (\nabla^\nu \bullet \boldsymbol{\Lambda})\, \boldsymbol{\Lambda} \bullet \mathbf{f}^\lambda dL + \left(\frac{\mathbf{n}^* \bullet \mathbf{f}}{\mathbf{e} \bullet \mathbf{n}^*}\right)\Bigg|_{\text{ends}} \tag{7.46}$$

C[2, (0, 0), 1] Curl

$$\int_L \nabla^s \times \mathbf{f} dL = \nabla^\nu \times \int_L (\mathbf{f}^\lambda + \mathbf{f}^n)\, dC + \int_L (\nabla^\nu \bullet \boldsymbol{\Lambda})\, \boldsymbol{\Lambda} \times \mathbf{f}^\nu dL + \left(\frac{\mathbf{n}^* \times \mathbf{f}}{\mathbf{e} \bullet \mathbf{n}^*}\right)\Bigg|_{\text{ends}} \tag{7.47}$$

T[2, (0, 0), 1] Transport

$$\int_L \frac{\partial f}{\partial t}\Bigg|_{\mathbf{u}} dL = \frac{\partial}{\partial t}\int_L f dL - \left(\frac{\mathbf{n}^* \bullet \mathbf{w} f}{\mathbf{e} \bullet \mathbf{n}^*}\right)\Bigg|_{\text{ends}} \tag{7.48}$$

7.3.5 Curvilineal Integration of Curvilineal Operations, [1, (0, 0), 1]

Figure and Notation

Refer to figure 7.4.

- C is the curve of integration,
- $\boldsymbol{\lambda}$ is the unit vector tangent to curve C,
- $\mathbf{e}$ is the unit vector tangent to curve C at its endpoints positive in the direction out from the curve,
- ∇^c is the curvilineal operator such that $\nabla^c = \boldsymbol{\lambda}\boldsymbol{\lambda} \bullet \nabla$,
- $\mathbf{w}$ is the velocity of C,
- $\mathbf{w}^c$ is the velocity of C tangent to C such that $\mathbf{w}^c = \boldsymbol{\lambda}\boldsymbol{\lambda} \bullet \mathbf{w}$,
- $\mathbf{w}^s$ is the velocity of C normal to C such that $\mathbf{w}^s = \mathbf{w} - \mathbf{w}^c$,
- f is a curvilineal function $f(\mathbf{l}, t)$,
- $\mathbf{f}$ is a curvilineal vector $\mathbf{f}(\mathbf{l}, t)$,
- $\mathbf{f}^c$ is a vector tangent to C such that $\mathbf{f}^c = \boldsymbol{\lambda}\boldsymbol{\lambda} \bullet \mathbf{f}$, and
- $\mathbf{f}^s$ is a vector normal to C such that $\mathbf{f}^s = \mathbf{f} - \mathbf{f}^c$.

G[1, (0, 0), 1] Gradient

$$\int_C \nabla^c f dC = -\int_C (\boldsymbol{\lambda}\bullet\nabla^c\boldsymbol{\lambda}) f dC + (\mathbf{e}f)|_{\text{ends}} \tag{7.49}$$

D[1, (0, 0), 1] Divergence

$$\int_C \nabla^c\bullet\mathbf{f} dC = -\int_C (\boldsymbol{\lambda}\bullet\nabla^c\boldsymbol{\lambda})\bullet\mathbf{f}^s dC + (\mathbf{e}\bullet\mathbf{f}^c)|_{\text{ends}} \tag{7.50}$$

C[1, (0, 0), 1] Curl

$$\int_C \nabla^c\times\mathbf{f} dC = -\int_C (\boldsymbol{\lambda}\bullet\nabla^c\boldsymbol{\lambda})\times\mathbf{f} dC + (\mathbf{e}\times\mathbf{f}^s)|_{\text{ends}} \tag{7.51}$$

T[1, (0, 0), 1] Transport

$$\int_C \left.\frac{\partial f}{\partial t}\right|_1 dC = \frac{d}{dt}\int_C f dC + \int_C (\boldsymbol{\lambda}\bullet\nabla^c\boldsymbol{\lambda})\bullet\mathbf{w}^s f dC - (\mathbf{e}\bullet\mathbf{w}^c f)|_{\text{ends}} \tag{7.52}$$

7.3.6 Integration of Curvilineal Operations over a Straight Line Fixed in Space, [1, (0, 0), 1]

Figure and Notation

Refer to figure 7.6 (or 7.7).

- L is the straight line segment of integration,
- $\boldsymbol{\Lambda}$ is the unit vector tangent to L,
- $\mathbf{e}$ is the unit vector tangent to L at its endpoints positive in the direction out from the line segment,
- ∇^c is the curvilineal operator such that $\nabla^c = \boldsymbol{\Lambda}\boldsymbol{\Lambda}\bullet\nabla$,
- $\mathbf{w}$ is the velocity of the endpoints of L such that, $\mathbf{w} = \boldsymbol{\Lambda}\boldsymbol{\Lambda}\bullet\mathbf{w}$
- f is a spatial function $f(\mathbf{x}, t)$,
- $\mathbf{f}$ is a spatial vector $\mathbf{f}(\mathbf{x}, t)$,
- $\mathbf{f}^c$ is a vector tangent to L such that $\mathbf{f}^c = \boldsymbol{\Lambda}\boldsymbol{\Lambda}\bullet\mathbf{f}$, and
- $\mathbf{f}^s$ is a vector normal to L such that $\mathbf{f}^s = \mathbf{f} - \mathbf{f}^c$.

G[1, (0, 0), 1] Gradient

$$\int_L \nabla^c f dL = (\mathbf{e}f)|_{\text{ends}} \tag{7.53}$$

D[1, (0, 0), 1] Divergence

$$\int_L \nabla^c \bullet \mathbf{f} dL = (\mathbf{e} \bullet \mathbf{f}^c)|_{\text{ends}} \tag{7.54}$$

C[1, (0, 0), 1] Curl

$$\int_L \nabla^c \times \mathbf{f} dL = (\mathbf{e} \times \mathbf{f}^s)|_{\text{ends}} \tag{7.55}$$

T[1, (0, 0), 1] Transport

$$\int_C \left.\frac{\partial f}{\partial t}\right|_1 dC = \frac{d}{dt}\int_C f dC - (\mathbf{e} \bullet \mathbf{w} f)|_{\text{ends}} \tag{7.56}$$

7.4 CONCLUSION

The integration theorems presented in this chapter are useful tools for transforming a balance equation from the microscale to the megascale. In fact, the theorems may be applied to affect a change from any continuum scale to the megascale.

The theorems most commonly used are those from the $[3, (0, 0), 3]$ family, equations (7.1) through (7.4). Included among these are the standard divergence theorem and the transport theorem typically encountered in undergraduate calculus courses, fluid mechanics, and continuum mechanics [e.g. *Kreyszig*, 1979; *Whitaker*, 1968; *Bowen*, 1989]. The family designated as $[3, (0, 0), 2]$ for integration over a flat plane, equations (7.17) through (7.20), are particularly useful for converting equations that describe three-dimensional processes to the forms appropriate for approximating the process as one-dimensional along a region of variable cross-section. The description of flow along a river channel is an example of a physical problem that can be described using these theorems. Similarly, the family designated as $[3, (0, 0), 1]$ for integration over a straight line, equations (7.37) through (7.40), are important to convert a fully three-dimensional equation to a two dimensional approximation. This approach is useful, for example, in obtaining the shallow water flow equations from the full equations for mass and momentum conservation. These types of transformations are typically made in a pseudo rigorous manner that can result in neglect of terms, such as the curvature of the earth, that may be important in some situations. The availability of the theorems, as provided here, assures that important terms will be neglected only if a conscious effort is made to do so. The theorems in this chapter will prove useful in a variety of fields and applications when it is desired to convert a balance equation at a small scale to one at a larger scale.

7.5 REFERENCES

Bowen, R. M., *Introduction to Continuum Mechanics for Engineers*, Plenum, New York, 1989.

Kreyszig, E., *Advanced Engineering Mathematics*, 4th ed., Wiley, New York, 1979.

Whitaker, S., *Introduction to Fluid Mechanics*, Prentice-Hall, Englewood Cliffs, 1968.

CHAPTER EIGHT

AVERAGING THEOREMS

8.0 INTRODUCTION

In this chapter, theorems are presented that change the scale of a differential operator by averaging. An averaging volume, also called a representative elementary volume (REV), may be located in space and then integration is performed over volumes, surfaces, or curves contained within the averaging volume. Because volumes are located at every position in space, the averages obtained have a functional dependence on spatial coordinates. The averaging theorems are particularly useful, for example, in problems of flow in porous media where it is desirable to change spatial scales from the pore scale (microscale) to the scale of a representative region in the system (macroscale) or the scale of the system itself (megascale). This chapter will merely present the theorems that have been derived using the generalized functions and explain the notation used, as appropriate. The notation for theorem identification has been given in Chapter 5, and useful identities and sample derivations appear in Chapter 6. Averaging theorems will be presented for four different operators: gradient (G), divergence (D), curl (C), and the time derivative (T).

The G, D, and C theorems are taken here to include those averaging theorems which relate an integral over a volume, surface, or curve within an averaging volume of the gradient, divergence, and curl of a function to the macroscopic gradient, divergence, and curl of the integral of the function, respectively. Thus integrals of microscopic derivatives are converted to combinations of appropriate macroscopic and megascopic counterparts. The gradient, divergence, and curl averaging theorems presented here comprise, respectively, the sets $G[i, (m, n), k]$, $D[i, (m, n), k]$, and $C[i, (m, n), k]$ where $i = 1, 2$, or 3, $m + n + k = 3$, and $m \times n = 0$ with all indices being non-negative. The most commonly encountered of these theorems is the set $[3, (3, 0), 0]$ which transforms a three-dimensional microscopic del operator to the three-dimensional macroscopic form. Note that in the defining expression for the averaging theorems - even when $i = 2$, indicating a surficial operator, or $i = 1$, indicating a curvilineal operator - transformation is such that the sum of macroscopic and megascopic coordinates is 3. For example, the theorem set $[1, (3, 0), 0]$ involves a change from differentiation of a function along the curves within an averaging volume to variation in space of the integral of that function over the curves.

Averaging theorems for the time derivative relate the integral of that derivative to the derivative of the integral. The theorems involve integration of the partial time derivative, evaluated holding spatial, surficial, or curvilineal coordinates constant, over a volume, surface, or curve contained within the averaging volume. Thus a time derivative of a function of microscopic coordinates is changed to a time derivative of a macroscopic-megascopic function. The time derivative averaging theorems are identified as $T[i, (m, n), k]$ where $i = 1, 2$, or 3, $m+n+k = 3$, and $m \times n = 0$ with all indices being non-negative.

Within the REV, different phases coexist at the microscale and occupy different portions of the REV. In the presentation of the theorems, when the first index is 3, the microscopic dimensionality is 3, indicating a spatial operator, or the partial time derivative in space, acts on a spatial property defined for the α-phase. For a microscopic dimensionality of 2, a surficial operator, or the partial time derivative fixed to a surface, acts on a surficial property of an interface between two phases. These phases are indicated as the α- and β-phases and the interface between them is the $\alpha\beta$-interface. When the microscopic dimensionality is 1, a curvilineal operator, or the partial time derivative at a point fixed to a curve, acts on a curvilineal property of a curve where three interfaces meet. If three phases are indicated as α, β, and ϵ, three interface types denoted as $\alpha\beta$, $\alpha\epsilon$, and $\beta\epsilon$ may be formed. The curve where these three interface types meet is referred to as the $\alpha\beta\epsilon$-intersect. Caution should be exercised to be sure the notation in each of the theorems below is understood as sometimes, for example, $S_{\alpha\beta}$ is used to indicate the interface between the α-phase and all other phases while at other times $S_{\alpha\beta}$ indicates only the interface between the α- and β-phases.

8.1 THREE MACROSCOPIC DIMENSIONS, [*i*, (3, 0), 0]

The averaging volume for this set of theorems is the spherical volume as depicted in figures 8.1 through 8.3. The radius of the sphere is independent of position and time. Averaging theorems in this section convert microscopic derivatives to three-dimensional macroscopic derivatives. Since the third index in the theorem in the theorem identifier is zero, no flux terms on the external boundary of the REV appear in the theorems.

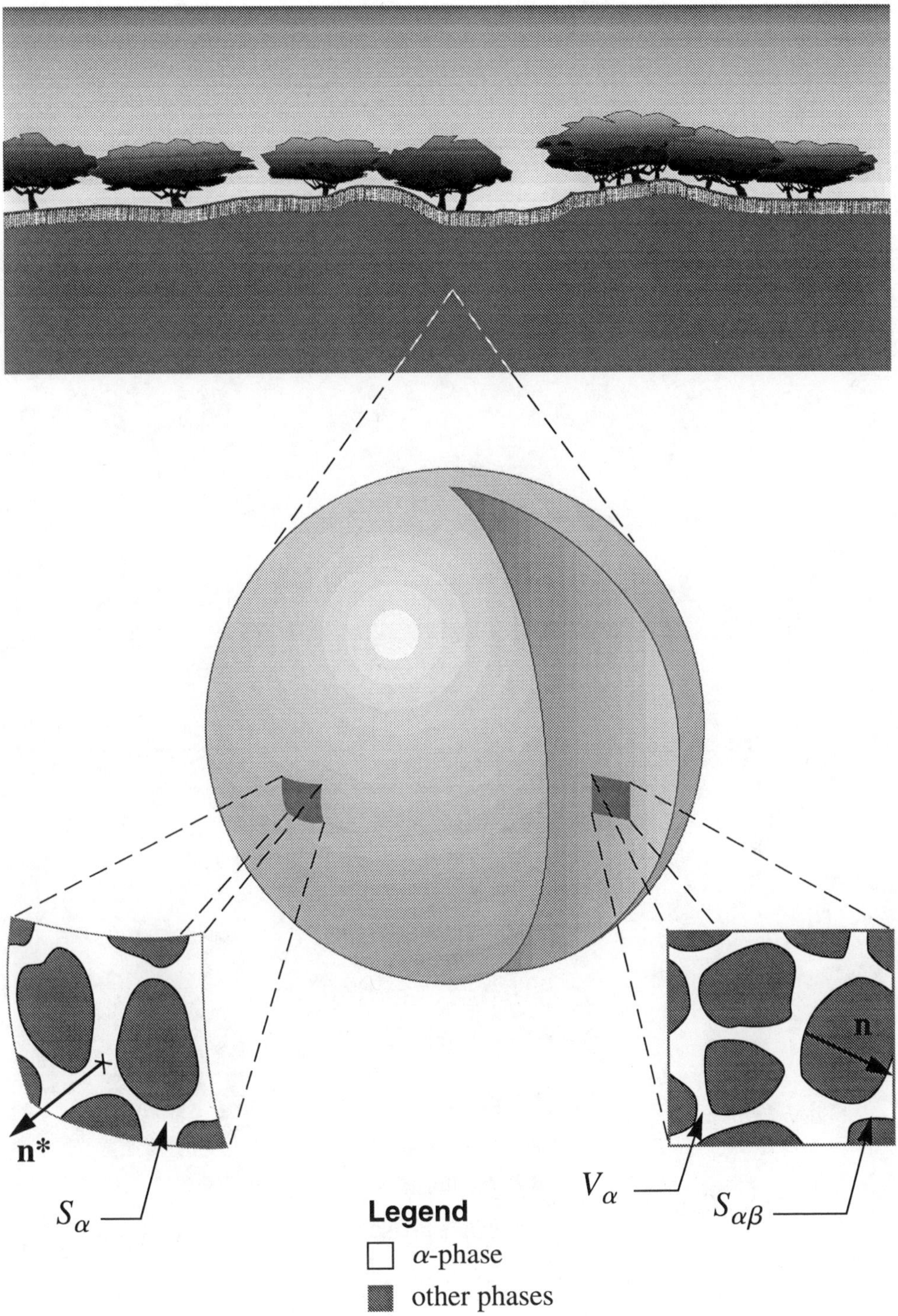

Figure 8.1. Spherical REV, independent of space and time (with wedge removed for illustrative purposes), used to develop averaging theorems for spatial operators, equations (8.1) through (8.4) and (8.13) through (8.16).

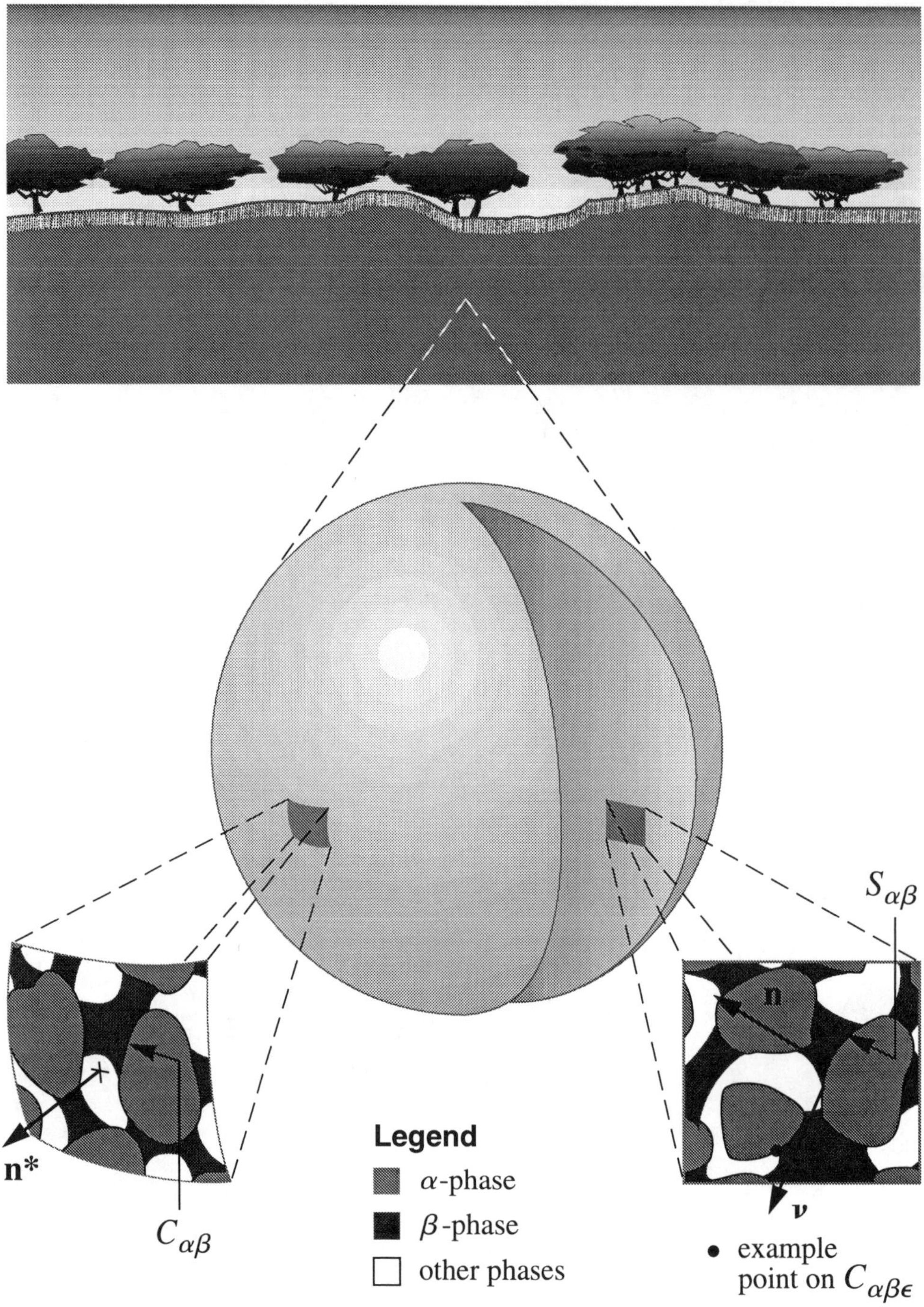

Figure 8.2. Spherical REV, independent of space and time (with wedge removed for illustrative purposes), used to develop averaging theorems for surficial operators, equations (8.5) through (8.8) and (8.17) through (8.20).

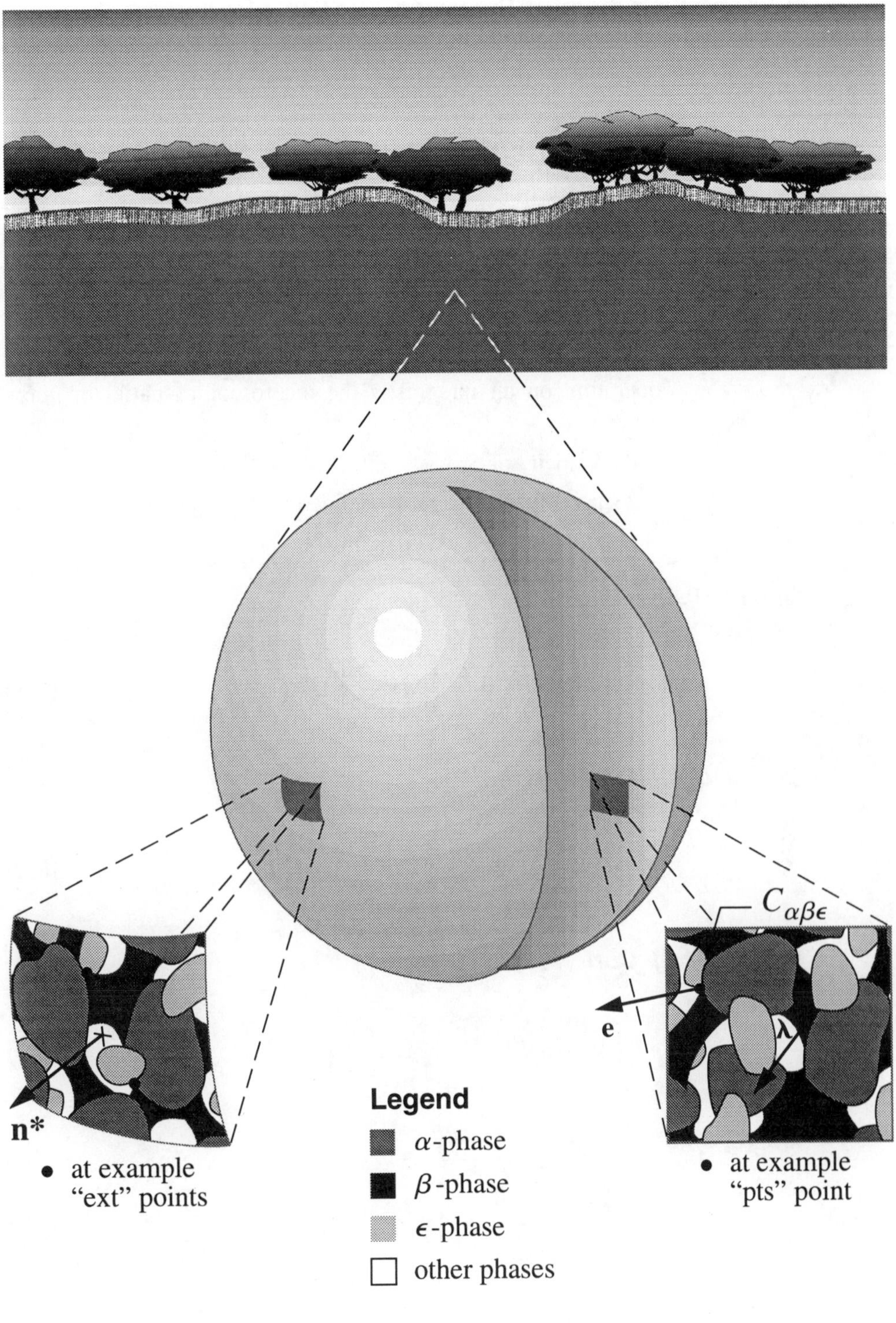

Figure 8.3. Spherical REV, independent of space and time (with wedge removed for illustrative purposes), used to develop averaging theorems for curvilineal operators, equations (8.9) through (8.12) and (8.21) through (8.24).

8.1.1 Spatial Operator Theorems, [3, (3, 0), 0]

Figure and Notation

Refer to figure 8.1

- V_α is the portion of the REV occupied by the α-phase,
- $S_{\alpha\beta}$ is the surface within the REV between the α-phase and all other phases,
- $\mathbf{n}$ is the unit vector normal to $S_{\alpha\beta}$ pointing outward from the α-phase,
- ∇ in the integrand is the microscopic spatial operator, $\nabla = \nabla_{\xi}$,
- ∇ operating on an integral is the macroscopic spatial operator, $\nabla = \nabla_{\mathbf{x}}$,
- $\mathbf{w}$ is the velocity of $S_{\alpha\beta}$,
- f is a spatial function $f(\mathbf{x}, t)$ defined in the α-phase, and
- $\mathbf{f}$ is a spatial vector $\mathbf{f}(\mathbf{x}, t)$ defined in the α-phase.

G[3, (3, 0), 0] Gradient

$$\int_{V_\alpha} \nabla f dV = \nabla \int_{V_\alpha} f dV + \int_{S_{\alpha\beta}} \mathbf{n} f dS \tag{8.1}$$

D[3, (3, 0), 0] Divergence

$$\int_{V_\alpha} \nabla \cdot \mathbf{f} dV = \nabla \cdot \int_{V_\alpha} \mathbf{f} dV + \int_{S_{\alpha\beta}} \mathbf{n} \cdot \mathbf{f} dS \tag{8.2}$$

C[3, (3, 0), 0] Curl

$$\int_{V_\alpha} \nabla \times \mathbf{f} dV = \nabla \times \int_{V_\alpha} \mathbf{f} dV + \int_{S_{\alpha\beta}} \mathbf{n} \times \mathbf{f} dS \tag{8.3}$$

T[3, (3, 0), 0] Transport

$$\int_{V_\alpha} \left.\frac{\partial f}{\partial t}\right|_{\mathbf{x}} dV = \frac{\partial}{\partial t} \int_{V_\alpha} f dV - \int_{S_{\alpha\beta}} \mathbf{n} \cdot \mathbf{w} f dS \tag{8.4}$$

8.1.2 Surficial Operator Theorems, [2, (3, 0), 0]

Figure and Notation

Refer to figure 8.2

$S_{\alpha\beta}$ is the surface within the REV between the α- and β-phases,

$C_{\alpha\beta\epsilon}$ is the boundary curve of $S_{\alpha\beta}$ within the REV that also is the location where the α- and β-phases meet other phases,

n is the unit vector normal to $S_{\alpha\beta}$,

$\boldsymbol{\nu}$ is the unit normal vector on the $C_{\alpha\beta\epsilon}$ curve positive outward from $S_{\alpha\beta}$ such that $\mathbf{n} \cdot \boldsymbol{\nu} = 0$,

∇^s is the microscopic surficial operator such that $\nabla^s = \nabla_\xi - \mathbf{nn} \cdot \nabla_\xi$,

∇ operating on an integral is the macroscopic spatial operator, $\nabla = \nabla_\mathbf{x}$,

w is the velocity of $S_{\alpha\beta}$,

$\mathbf{w}^c$ is the velocity of $S_{\alpha\beta}$ normal to $S_{\alpha\beta}$, $\mathbf{w}^c = \mathbf{nn} \cdot \mathbf{w}$,

$\mathbf{w}^s$ is the velocity of $S_{\alpha\beta}$ tangent to $S_{\alpha\beta}$, $\mathbf{w}^s = \mathbf{w} - \mathbf{w}^c$,

f is a surficial function $f(\mathbf{u}, t)$ defined on $S_{\alpha\beta}$,

f is a surficial vector $\mathbf{f}(\mathbf{u}, t)$ defined on $S_{\alpha\beta}$, and

$\mathbf{f}^s$ is a vector tangent to $S_{\alpha\beta}$ such that $\mathbf{f}^s = \mathbf{f} - \mathbf{nn} \cdot \mathbf{f}$.

G[2, (3, 0), 0] Gradient

$$\int_{S_{\alpha\beta}} \nabla^s f dS = \nabla \int_{S_{\alpha\beta}} f dS - \nabla \cdot \int_{S_{\alpha\beta}} \mathbf{nn} f dS + \int_{S_{\alpha\beta}} (\nabla^s \cdot \mathbf{n}) \, \mathbf{n} f dS + \int_{C_{\alpha\beta\epsilon}} \boldsymbol{\nu} f dC \quad \textbf{(8.5)}$$

D[2, (3, 0), 0] Divergence

$$\int_{S_{\alpha\beta}} \nabla^s \cdot \mathbf{f} dS = \nabla \cdot \int_{S_{\alpha\beta}} \mathbf{f}^s dS + \int_{S_{\alpha\beta}} (\nabla^s \cdot \mathbf{n}) \, \mathbf{n} \cdot \mathbf{f} dS + \int_{C_{\alpha\beta\epsilon}} \boldsymbol{\nu} \cdot \mathbf{f}^s dC \quad \textbf{(8.6)}$$

C[2, (3, 0), 0] Curl

$$\int_{S_{\alpha\beta}} \nabla^s \times \mathbf{f} dS = \nabla \times \int_{S_{\alpha\beta}} \mathbf{f} dS - \nabla \cdot \int_{S_{\alpha\beta}} \mathbf{nn} \times \mathbf{f}^s dS$$

$$+ \int_{S_{\alpha\beta}} (\nabla^s \cdot \mathbf{n}) \, \mathbf{n} \times \mathbf{f}^s dS + \int_{C_{\alpha\beta\epsilon}} \boldsymbol{\nu} \times \mathbf{f} dC \quad \textbf{(8.7)}$$

T[2, (3, 0), 0] Transport

$$\int_{S_{\alpha\beta}} \left.\frac{\partial f}{\partial t}\right|_{\mathbf{u}} dS = \frac{\partial}{\partial t}\int_{S_{\alpha\beta}} f dS + \nabla\bullet \int_{S_{\alpha\beta}} \mathbf{w}^c f dS - \int_{S_{\alpha\beta}} (\nabla^s\bullet\mathbf{n})\,\mathbf{n}\bullet\mathbf{w}^c f dS - \int_{C_{\alpha\beta\epsilon}} \boldsymbol{\nu}\bullet\mathbf{w}^s f dC \tag{8.8}$$

8.1.3 Curvilineal Operator Theorems, [1, (3, 0), 0]

Figure and Notation

Refer to figure 8.3

- $C_{\alpha\beta\epsilon}$ is the contact curve within the REV that is the location where the α, β, and ϵ phases meet,
- pts refers to the end points of the $C_{\alpha\beta\epsilon}$ curve within the REV where the α, β, and ϵ phases meet a fourth phase,
- $\boldsymbol{\lambda}$ is the unit vector tangent to curve $C_{\alpha\beta\epsilon}$,
- $\mathbf{e}$ is the unit vector tangent to curve $C_{\alpha\beta\epsilon}$ at its endpoints within the REV positive in the direction out from the curve such that $\boldsymbol{\lambda}\bullet\mathbf{e} = \pm 1$,
- ∇^c is the microscopic curvilineal operator such that $\nabla^c = \boldsymbol{\lambda}\boldsymbol{\lambda}\bullet\nabla_{\xi}$,
- ∇ outside the integral is the macroscopic spatial operator, $\nabla = \nabla_{\mathbf{x}}$,
- $\mathbf{w}$ is the velocity of $C_{\alpha\beta\epsilon}$,
- $\mathbf{w}^c$ is the velocity of $C_{\alpha\beta\epsilon}$ tangent to $C_{\alpha\beta\epsilon}$, $\mathbf{w}^c = \boldsymbol{\lambda}\boldsymbol{\lambda}\bullet\mathbf{w}$,
- $\mathbf{w}^s$ is the velocity of $C_{\alpha\beta\epsilon}$ normal to $C_{\alpha\beta\epsilon}$, $\mathbf{w}^s = \mathbf{w} - \mathbf{w}^c$,
- f is a curvilineal function $f(\mathbf{l}, t)$ defined on $C_{\alpha\beta\epsilon}$,
- $\mathbf{f}$ is a curvilineal vector $\mathbf{f}(\mathbf{l}, t)$ defined on $C_{\alpha\beta\epsilon}$,
- $\mathbf{f}^c$ is the vector tangent to $C_{\alpha\beta\epsilon}$ such that $\mathbf{f}^c = \boldsymbol{\lambda}\boldsymbol{\lambda}\bullet\mathbf{f}$, and
- $\mathbf{f}^s$ is the vector normal to $C_{\alpha\beta\epsilon}$ such that $\mathbf{f}^s = \mathbf{f} - \mathbf{f}^c$.

G[1, (3, 0), 0] Gradient

$$\int_{C_{\alpha\beta\epsilon}} \nabla^c f dC = \nabla\bullet \int_{C_{\alpha\beta\epsilon}} \boldsymbol{\lambda}\boldsymbol{\lambda} f dC - \int_{C_{\alpha\beta\epsilon}} \boldsymbol{\lambda}\bullet\nabla^c\boldsymbol{\lambda} f dC + \sum_{\text{pts}} \mathbf{e} f \tag{8.9}$$

D[1, (3, 0), 0] Divergence

$$\int_{C_{\alpha\beta\epsilon}} \nabla^c\bullet\mathbf{f} dC = \nabla\bullet \int_{C_{\alpha\beta\epsilon}} \mathbf{f}^c dC - \int_{C_{\alpha\beta\epsilon}} \boldsymbol{\lambda}\bullet\nabla^c\boldsymbol{\lambda}\bullet\mathbf{f}^s + \sum_{\text{pts}} \mathbf{e}\bullet\mathbf{f}^c \tag{8.10}$$

C[1, (3, 0), 0] Curl

$$\int_{C_{\alpha\beta\epsilon}} \nabla^c \times \mathbf{f} dC = \nabla \cdot \int_{C_{\alpha\beta\epsilon}} \boldsymbol{\lambda}\boldsymbol{\lambda} \times \mathbf{f}^s dC - \int_{C_{\alpha\beta\epsilon}} (\boldsymbol{\lambda} \cdot \nabla^c \boldsymbol{\lambda}) \times \mathbf{f} dC + \sum_{\text{pts}} \mathbf{e} \times \mathbf{f}^s \quad (8.11)$$

T[1, (3, 0), 0] Transport

$$\int_{C_{\alpha\beta\epsilon}} \left.\frac{\partial f}{\partial t}\right|_1 dC = \frac{\partial}{\partial t} \int_{C_{\alpha\beta\epsilon}} f dC + \nabla \cdot \int_{C_{\alpha\beta\epsilon}} \mathbf{w}^s f dC + \int_{C_{\alpha\beta\epsilon}} \boldsymbol{\lambda} \cdot \nabla^c \boldsymbol{\lambda} \cdot \mathbf{w}^s f dC - \sum_{\text{pts}} \mathbf{e} \cdot \mathbf{w}^c f$$

(8.12)

8.2 THREE MACROSCOPIC DIMENSIONS, [*i*, (0, 3), 0]

As in Section 8.1, the averaging volume for this set of theorems is the spherical volume with constant radius as shown in figures 8.1 through 8.3. In contrast to the theorems given in the previous section, however, the theorems in Section 8.2 will not contain macroscopic spatial derivatives but will be formulated in terms of fluxes on the boundary of the REV. This difference is indicated by the interchange of the second and third indices in the designation of the theorem family.

8.2.1 Spatial Operator Theorems, [3, (0, 3), 0]

Figure and Notation

Refer to figure 8.1

- V_α is the portion of the REV occupied by the α-phase,
- S_α is the portion of the external boundary of the REV that intersects the α-phase,
- $S_{\alpha\beta}$ is the surface within the REV between the α-phase and all other phases,
- $\mathbf{n}^*$ is the unit vector normal to the external boundary of the REV pointing outward from the REV,
- $\mathbf{n}$ is the unit vector normal to $S_{\alpha\beta}$ pointing outward from the α-phase,
- ∇ is the microscopic spatial operator, $\nabla = \nabla_\xi$,
- $\mathbf{w}$ is the velocity of $S_{\alpha\beta}$,
- f is a spatial function $f(\mathbf{x}, t)$ defined in the α-phase, and
- $\mathbf{f}$ is a spatial vector $\mathbf{f}(\mathbf{x}, t)$ defined in the α-phase.

G[3, (0, 3), 0] Gradient

$$\int_{V_\alpha} \nabla f dV = \int_{S_\alpha} \mathbf{n}^* f dS + \int_{S_{\alpha\beta}} \mathbf{n} f dS \quad (8.13)$$

D[3, (0, 3), 0] Divergence

$$\int_{V_\alpha} \nabla \bullet \mathbf{f} dV = \int_{S_\alpha} \mathbf{n}^* \bullet \mathbf{f} dS + \int_{S_{\alpha\beta}} \mathbf{n} \bullet \mathbf{f} dS \quad (8.14)$$

C[3, (0, 3), 0] Curl

$$\int_{V_\alpha} \nabla \times \mathbf{f} dV = \int_{S_\alpha} \mathbf{n}^* \times \mathbf{f} dS + \int_{S_{\alpha\beta}} \mathbf{n} \times \mathbf{f} dS \quad (8.15)$$

T[3, (0, 3), 0] Transport

$$\int_{V_\alpha} \left.\frac{\partial f}{\partial t}\right|_{\mathbf{x}} dV = \frac{\partial}{\partial t} \int_{V_\alpha} f dV - \int_{S_{\alpha\beta}} \mathbf{n} \bullet \mathbf{w} f dS \quad (8.16)$$

8.2.2 Surficial Operator Theorems, [2, (0, 3), 0]

Figure and Notation

Refer to figure 8.2

- $S_{\alpha\beta}$ is the surface within the REV between the α- and β-phases,
- $C_{\alpha\beta}$ is the curve of intersection of the $S_{\alpha\beta}$ surface with the external surface of the REV,
- $C_{\alpha\beta\epsilon}$ is the boundary curve of $S_{\alpha\beta}$ within the REV that also is the location where the α- and β- phases meet a third phase,
- $\mathbf{n}^*$ is the unit vector normal to the external boundary of the REV pointing outward from the REV,
- $\mathbf{n}$ is the unit vector normal to $S_{\alpha\beta}$,
- $\boldsymbol{\nu}$ is the unit normal vector on the $C_{\alpha\beta\epsilon}$ and $C_{\alpha\beta}$ curves positive outward from $S_{\alpha\beta}$ such that $\mathbf{n} \bullet \boldsymbol{\nu} = 0$,
- ∇^s is the microscopic surficial operator such that $\nabla^s = \nabla_\xi - \mathbf{n}\mathbf{n} \bullet \nabla_\xi$,
- $\mathbf{w}$ is the velocity of $S_{\alpha\beta}$,
- $\mathbf{w}^C$ is the velocity of $S_{\alpha\beta}$ normal to $S_{\alpha\beta}$, $\mathbf{w}^C = \mathbf{n}\mathbf{n} \bullet \mathbf{w}$,

$\mathbf{w}^s$ is the velocity of $S_{\alpha\beta}$ tangent to $S_{\alpha\beta}$, $\mathbf{w}^s = \mathbf{w} - \mathbf{w}^c$,
f is a surficial function $f(\mathbf{u}, t)$ defined on $S_{\alpha\beta}$,
$\mathbf{f}$ is a surficial vector $\mathbf{f}(\mathbf{u}, t)$ defined on $S_{\alpha\beta}$,
$\mathbf{f}^c$ is the vector normal to $S_{\alpha\beta}$ such that $\mathbf{f}^c = \mathbf{nn}\cdot\mathbf{f}$, and
$\mathbf{f}^s$ is the vector tangent to $S_{\alpha\beta}$ such that $\mathbf{f}^s = \mathbf{f} - \mathbf{f}^c$.

G[2, (0, 3), 0] Gradient

$$\int_{S_{\alpha\beta}} \nabla^s f dS = \int_{S_{\alpha\beta}} (\nabla^s\cdot\mathbf{n})\,\mathbf{n} f dS + \int_{C_{\alpha\beta}} \boldsymbol{\nu} f dC + \int_{C_{\alpha\beta\epsilon}} \boldsymbol{\nu} f dC \quad \textbf{(8.17)}$$

D[2, (0, 3), 0] Divergence

$$\int_{S_{\alpha\beta}} \nabla^s\cdot\mathbf{f} dS = \int_{S_{\alpha\beta}} (\nabla^s\cdot\mathbf{n})\,\mathbf{n}\cdot\mathbf{f}^c dS + \int_{C_{\alpha\beta}} \boldsymbol{\nu}\cdot\mathbf{f}^s dC + \int_{C_{\alpha\beta\epsilon}} \boldsymbol{\nu}\cdot\mathbf{f}^s dC \quad \textbf{(8.18)}$$

C[2, (0, 3), 0] Curl

$$\int_{S_{\alpha\beta}} \nabla^s \times \mathbf{f} dS = \int_{S_{\alpha\beta}} (\nabla^s\cdot\mathbf{n})\,\mathbf{n}\times\mathbf{f}^s dS + \int_{C_{\alpha\beta}} \boldsymbol{\nu}\times\mathbf{f} dC + \int_{C_{\alpha\beta\epsilon}} \boldsymbol{\nu}\times\mathbf{f} dC \quad \textbf{(8.19)}$$

T[2, (0, 3), 0] Transport

$$\int_{S_{\alpha\beta}} \left.\frac{\partial f}{\partial t}\right|_{\mathbf{u}} dS = \frac{\partial}{\partial t}\int_{S_{\alpha\beta}} f dS - \int_{S_{\alpha\beta}} (\nabla^s\cdot\mathbf{n})\,\mathbf{n}\cdot\mathbf{w}^c f dS + \int_{C_{\alpha\beta}} \frac{\mathbf{n}^*\cdot\mathbf{w} f}{\boldsymbol{\nu}\cdot\mathbf{n}^*} dC$$

$$- \int_{C_{\alpha\beta}} \boldsymbol{\nu}\cdot\mathbf{w}^s f dC - \int_{C_{\alpha\beta\epsilon}} \boldsymbol{\nu}\cdot\mathbf{w}^s f dC \quad \textbf{(8.20)}$$

8.2.3 Curvilineal Operator Theorems, [1, (0, 3), 0]

Figure and Notation

Refer to figure 8.3

$C_{\alpha\beta\epsilon}$ is the contact curve within the REV that is the location where the α, β, and ϵ phases meet,

ext refers to the points of intersection of the $C_{\alpha\beta\epsilon}$ curve with the external surface of the REV,

pts refers to the end points of the $C_{\alpha\beta\epsilon}$ curve within the REV where the α, β, and ϵ phases meet a fourth phase,

$\mathbf{n}^*$ is the unit vector normal to the external boundary of the REV pointing outward from the REV,

$\boldsymbol{\lambda}$ is the unit vector tangent to curve $C_{\alpha\beta\epsilon}$,

$\mathbf{e}$ is the unit vector tangent to curve $C_{\alpha\beta\epsilon}$ at its endpoints within the REV positive in the direction out from the curve such that $\boldsymbol{\lambda}\cdot\mathbf{e} = \pm 1$,

∇^c is the microscopic curvilineal operator such that $\nabla^c = \boldsymbol{\lambda}\boldsymbol{\lambda}\cdot\nabla_\xi$,

$\mathbf{w}$ is the velocity of $C_{\alpha\beta\epsilon}$,

$\mathbf{w}^c$ is the velocity of $C_{\alpha\beta\epsilon}$ tangent to $C_{\alpha\beta\epsilon}$ such that $\mathbf{w}^c = \boldsymbol{\lambda}\boldsymbol{\lambda}\cdot\mathbf{w}$,

$\mathbf{w}^s$ is the velocity of $C_{\alpha\beta\epsilon}$ normal to $C_{\alpha\beta\epsilon}$ such that $\mathbf{w}^s = \mathbf{w} - \mathbf{w}^c$,

f is a curvilineal function $f(\mathbf{l}, t)$ defined on $C_{\alpha\beta\epsilon}$,

$\mathbf{f}$ is a curvilineal vector $\mathbf{f}(\mathbf{l}, t)$ defined on $C_{\alpha\beta\epsilon}$,

$\mathbf{f}^c$ is the vector tangent to $C_{\alpha\beta\epsilon}$ such that $\mathbf{f}^c = \boldsymbol{\lambda}\boldsymbol{\lambda}\cdot\mathbf{f}$, and

$\mathbf{f}^s$ is the vector normal to $C_{\alpha\beta\epsilon}$ such that $\mathbf{f}^s = \mathbf{f} - \mathbf{f}^c$.

G[1, (0, 3), 0] Gradient

$$\int_{C_{\alpha\beta\epsilon}} \nabla^c f dC = -\int_{C_{\alpha\beta\epsilon}} (\boldsymbol{\lambda}\cdot\nabla^c\boldsymbol{\lambda}) f dC + \sum_{\text{ext}} \mathbf{e}f + \sum_{\text{pts}} \mathbf{e}f \tag{8.21}$$

D[1, (0, 3), 0] Divergence

$$\int_{C_{\alpha\beta\epsilon}} \nabla^c\cdot\mathbf{f} dC = -\int_{C_{\alpha\beta\epsilon}} (\boldsymbol{\lambda}\cdot\nabla^c\boldsymbol{\lambda})\cdot\mathbf{f}^s dC + \sum_{\text{ext}} \mathbf{e}\cdot\mathbf{f}^c + \sum_{\text{pts}} \mathbf{e}\cdot\mathbf{f}^c \tag{8.22}$$

C[1, (0, 3), 0] Curl

$$\int_{C_{\alpha\beta\epsilon}} \nabla^c\times\mathbf{f} dC = -\int_{C_{\alpha\beta\epsilon}} (\boldsymbol{\lambda}\cdot\nabla^c\boldsymbol{\lambda})\times\mathbf{f} dC + \sum_{\text{ext}} \mathbf{e}\times\mathbf{f}^s + \sum_{\text{pts}} \mathbf{e}\times\mathbf{f}^s \tag{8.23}$$

T[1, (0, 3), 0] Transport

$$\int_{C_{\alpha\beta\epsilon}} \left.\frac{\partial f}{\partial t}\right|_{\mathbf{l}} dC = \frac{\partial}{\partial t}\int_{C_{\alpha\beta\epsilon}} f dC + \int_{C_{\alpha\beta\epsilon}} \boldsymbol{\lambda}\cdot\nabla^c\boldsymbol{\lambda}\cdot\mathbf{w}^s f dC + \sum_{\text{ext}} \frac{\mathbf{n}^*\cdot\mathbf{w}f}{\mathbf{e}\cdot\mathbf{n}^*}$$

$$-\sum_{\text{ext}} \mathbf{e}\cdot\mathbf{w}^c f - \sum_{\text{pts}} \mathbf{e}\cdot\mathbf{w}^c f \tag{8.24}$$

8.3 TWO MACROSCOPIC, ONE MEGASCOPIC DIMENSION, [*i*, (2, 0), 1]

The averaging volume, or REV, for this set of averaging theorems is a circular cylinder with diameter D as given in figures 8.4 through 8.6. The radius of the cylinder is independent of position and time. The orientation of the axis of the cylinder is taken to be independent of time but is allowed to vary with position. However, this variation with position must be such that the radius of curvature of the surface orthogonal to the axis of the cylinder as a function of position must be much less than the diameter of the cylinder such that $|\nabla^S \cdot \mathbf{\Lambda}| D \ll 1$. The volume of the REV may change with time and position due to changes in its length as determined by the intersection of the cylinder with megascopic boundaries of the system. The theorems presented in this section will contain macroscopic derivatives in the two directions perpendicular to the axis of the cylindrical REV and flux terms over the cylinder ends where the REV intersects the megascopic system boundaries.

8.3.1 Spatial Operator Theorems, [3, (2, 0), 1]

Figure and Notation

Refer to figure 8.4

V_α is the portion of the REV occupied by the α-phase,

$S_{\alpha\beta}$ is the surface within the REV between the α-phase and all other phases,

$S_{\alpha_{\text{ends}}}$ is the portion of the end boundaries of the REV that is occupied by the α-phase,

$\mathbf{n}^*$ is the unit vector normal to the external boundary of the REV pointing outward from the REV,

$\mathbf{n}$ is the unit vector normal to the $S_{\alpha\beta}$ surface pointing outward from the α-phase,

$\mathbf{\Lambda}$ is the unit vector tangent to the axis of the REV,

∇ is the microscopic spatial operator, $\nabla = \nabla_{\boldsymbol{\xi}}$,

∇^S is the two-dimensional macroscopic operator in the directions orthogonal to the axis of the REV, $\nabla^S = \nabla_{\mathbf{x}} - \mathbf{\Lambda}\mathbf{\Lambda} \cdot \nabla_{\mathbf{x}}$,

$\mathbf{w}$ is the velocity of either $S_{\alpha\beta}$ or $S_{\alpha_{\text{ends}}}$,

f is a spatial function $f(\mathbf{x}, t)$ defined in the α-phase,

$\mathbf{f}$ is a spatial vector $\mathbf{f}(\mathbf{x}, t)$ defined in the α-phase, and

$\mathbf{f}^c$ is the normal to $S_{\alpha\beta}$ such that $\mathbf{f}^c = \mathbf{n}\mathbf{n} \cdot \mathbf{f}$.

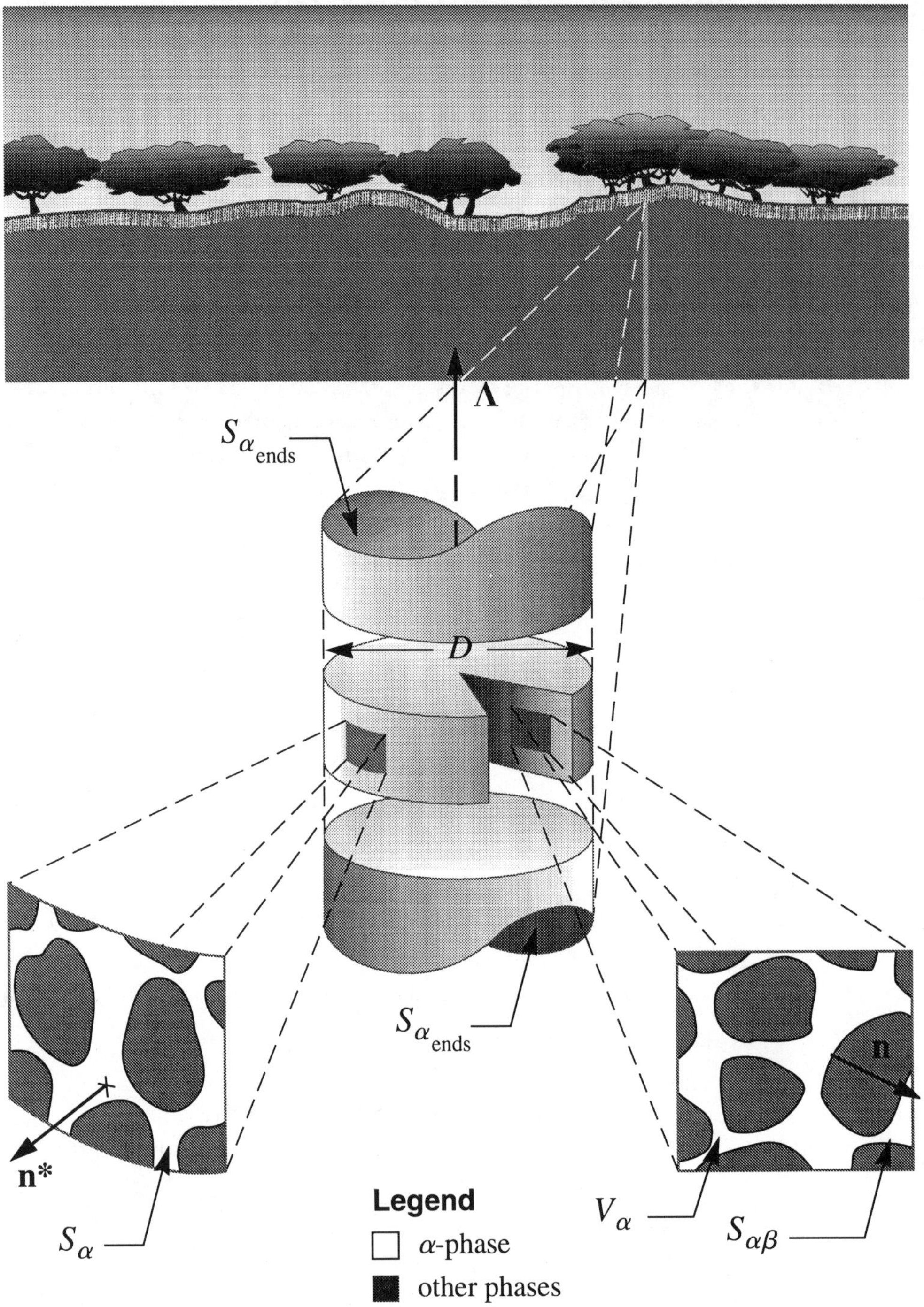

Figure 8.4. Cylindrical REV with constant radius but length and orientation dependent on space and time (wedge removed for illustrative purposes), used to develop averaging theorems for spatial operators, equations (8.25) through (8.28) and (8.37) through (8.40).

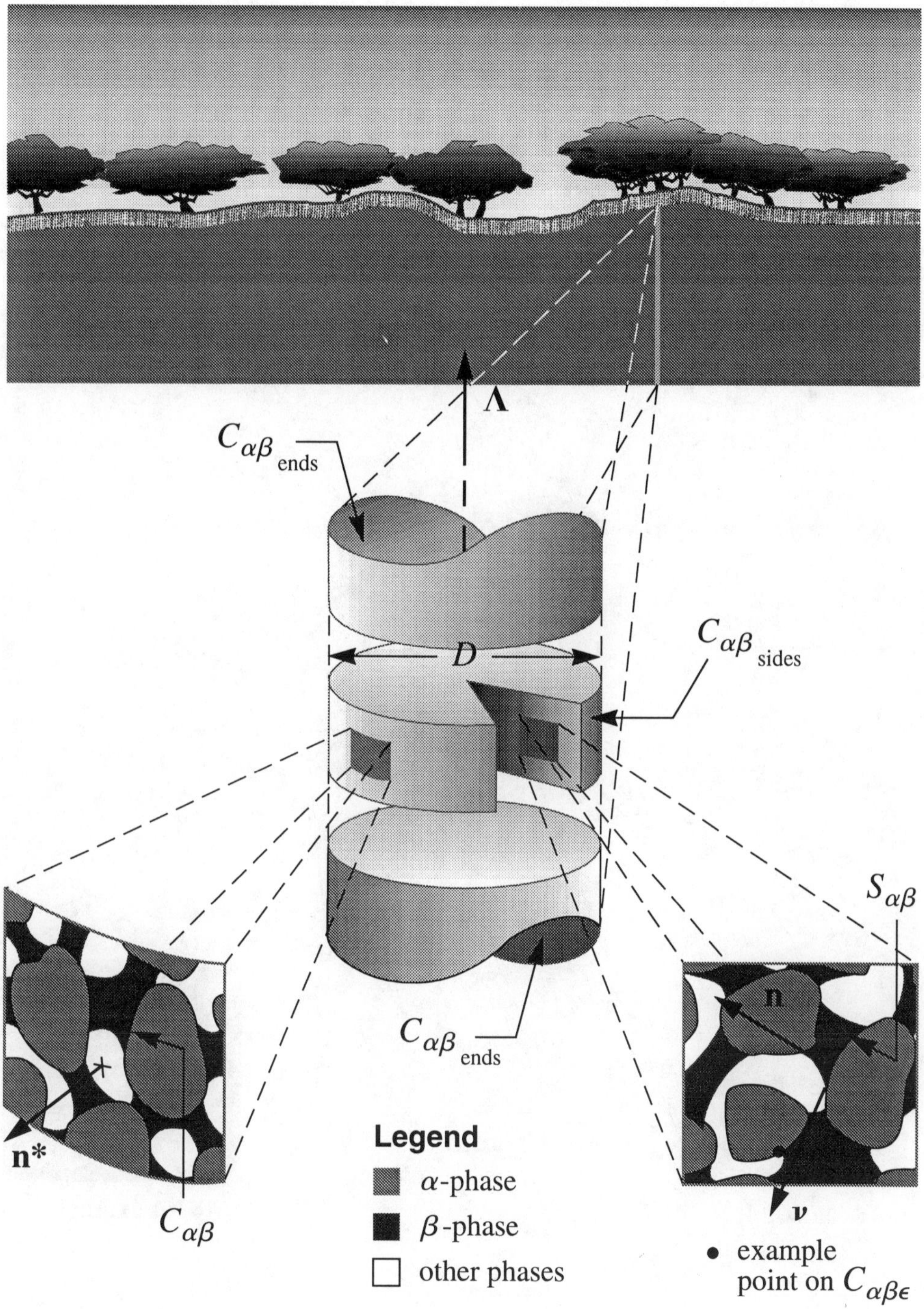

Figure 8.5. Cylindrical REV with constant radius but length and orientation dependent on space and time (wedge removed for illustrative purposes), used to develop averaging theorems for surficial operators, equations (8.29) through (8.32) and (8.41) through (8.44).

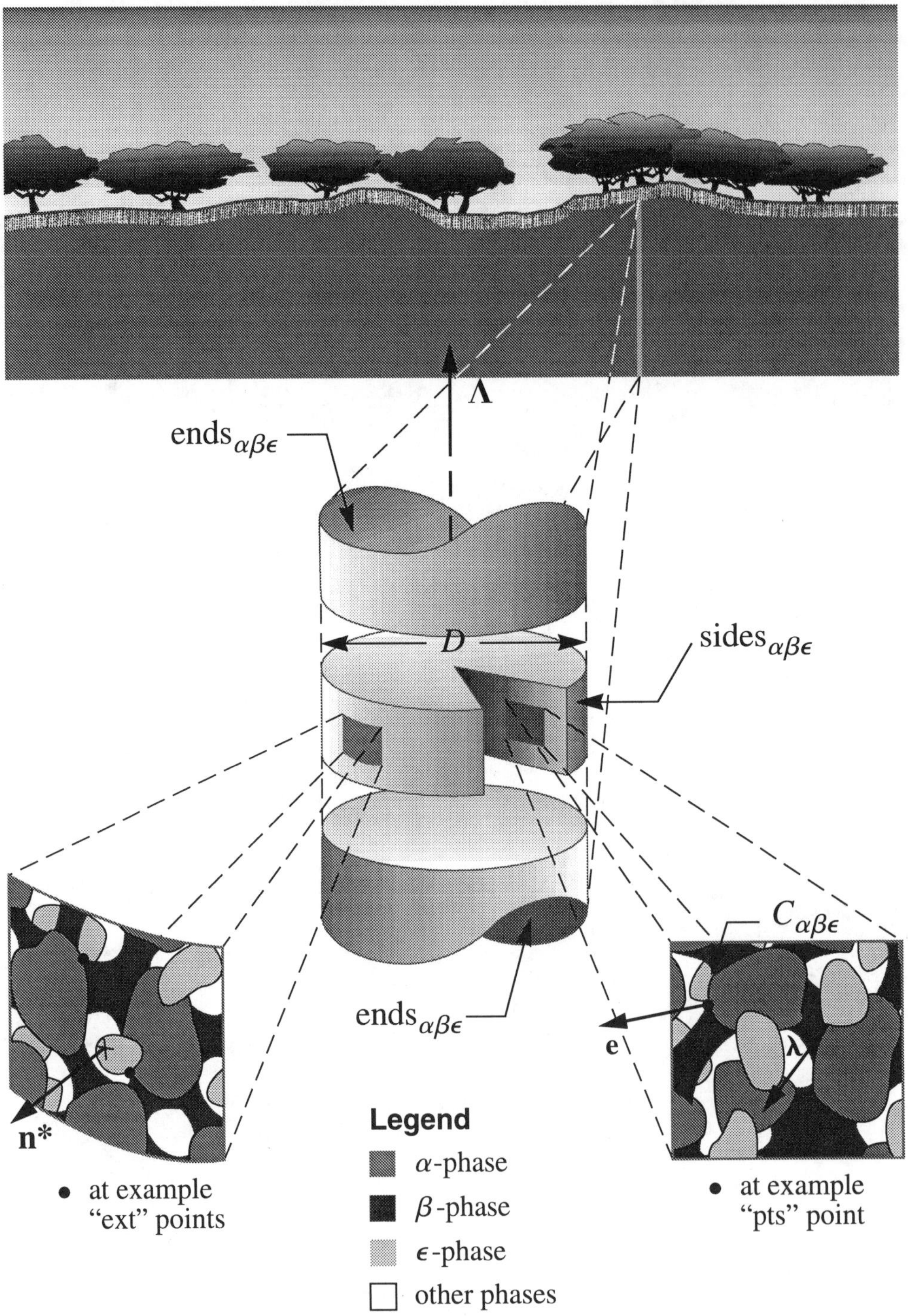

Figure 8.6. Cylindrical REV with constant radius but length and orientation dependent on space and time (wedge removed for illustrative purposes), used to develop averaging theorems for curvilineal operators, equations (8.33) through (8.36) and (8.45) through (8.48).

G[3, (2, 0), 1] Gradient

$$\int_{V_\alpha} \nabla f dV = \nabla^S \int_{V_\alpha} f dV + \int_{S_{\alpha\beta}} \mathbf{n} f dS + \int_{S_{\alpha_{\text{ends}}}} \mathbf{n}^* f dS \tag{8.25}$$

D[3, (2, 0), 1] Divergence

$$\int_{V_\alpha} \nabla \bullet \mathbf{f} dV = \nabla^S \bullet \int_{V_\alpha} \mathbf{f} dV + \int_{S_{\alpha\beta}} \mathbf{n} \bullet \mathbf{f}^c dS + \int_{S_{\alpha_{\text{ends}}}} \mathbf{n}^* \bullet \mathbf{f} dS \tag{8.26}$$

C[3, (2, 0), 1] Curl

$$\int_{V_\alpha} \nabla \times \mathbf{f} dV = \nabla^S \times \int_{V_\alpha} \mathbf{f} dV + \int_{S_{\alpha\beta}} \mathbf{n} \times \mathbf{f} dS + \int_{S_{\alpha_{\text{ends}}}} \mathbf{n}^* \times \mathbf{f} dS \tag{8.27}$$

T[3, (2, 0), 1] Transport

$$\int_{V_\alpha} \left.\frac{\partial f}{\partial t}\right|_{\mathbf{x}} dV = \frac{\partial}{\partial t} \int_{V_\alpha} f dV - \int_{S_{\alpha\beta}} \mathbf{n} \bullet \mathbf{w} f dS - \int_{S_{\alpha_{\text{ends}}}} \mathbf{n}^* \bullet \mathbf{w} f dS \tag{8.28}$$

8.3.2 Surficial Operator Theorems, [2, (2, 0), 1]

Figure and Notation

Refer to figure 8.5

$S_{\alpha\beta}$ is the surface between the α- and the β-phases,

$C_{\alpha\beta\epsilon}$ is the boundary curve of $S_{\alpha\beta}$ within the REV that also is the location where the α- and β-phases meet a third phase,

$C_{\alpha\beta_{\text{ends}}}$ is the curve at the ends of the REV formed by the intersection of $S_{\alpha\beta}$ with the ends of the cylinder,

$\mathbf{n}^*$ is the unit vector normal to the external boundary of the REV pointing outward from the REV,

$\mathbf{n}$ is the unit vector normal to the $S_{\alpha\beta}$ surface,

$\boldsymbol{\nu}$ is the unit vector tangent to the $S_{\alpha\beta}$ surface and normal to $C_{\alpha\beta\epsilon}$ and $C_{\alpha\beta_{\text{ends}}}$ such that $\mathbf{n} \bullet \boldsymbol{\nu} = 0$,

$\boldsymbol{\Lambda}$ is the unit vector tangent to the axis of the REV,

∇ is the microscopic spatial operator, $\nabla = \nabla_\xi$,

∇^s is the microscopic surficial operator, $\nabla^s = \nabla_\xi - \mathbf{nn} \bullet \nabla_\xi$,

∇^S is the two-dimensional macroscopic spatial operator in the directions orthogonal to the axis of the REV, $\nabla^S = \nabla_{\mathbf{x}} - \boldsymbol{\Lambda\Lambda} \bullet \nabla_{\mathbf{x}}$,

$\mathbf{w}$ is the velocity of $S_{\alpha\beta}$,
$\mathbf{w}^c$ is the velocity of $S_{\alpha\beta}$ normal to $S_{\alpha\beta}$, $\mathbf{w}^c = \mathbf{nn}\cdot\mathbf{w}$,
$\mathbf{w}^s$ is the velocity of $S_{\alpha\beta}$ tangent to $S_{\alpha\beta}$, $\mathbf{w}^s = \mathbf{w} - \mathbf{w}^c$,
f is a surficial function $f(\mathbf{u}, t)$ defined on $S_{\alpha\beta}$,
$\mathbf{f}$ is a surficial vector $\mathbf{f}(\mathbf{u}, t)$ defined on $S_{\alpha\beta}$,
$\mathbf{f}^c$ is the vector normal to $S_{\alpha\beta}$ such that $\mathbf{f}^c = \mathbf{nn}\cdot\mathbf{f}$, and
$\mathbf{f}^s$ is the vector tangent to $S_{\alpha\beta}$ such that $\mathbf{f}^s = \mathbf{f} - \mathbf{f}^c$.

G[2, (2, 0), 1] Gradient

$$\int_{S_{\alpha\beta}} \nabla^s f dS = \nabla^S \int_{S_{\alpha\beta}} f dS - \nabla^S\cdot \int_{S_{\alpha\beta}} \mathbf{nn} f dS + \int_{S_{\alpha\beta}} (\nabla^s\cdot\mathbf{n})\,\mathbf{n} f dS + \int_{C_{\alpha\beta\epsilon}} \boldsymbol{\nu} f dC + \int_{C_{\alpha\beta_{\text{ends}}}} \boldsymbol{\nu} f dC \tag{8.29}$$

D[2, (2, 0), 1] Divergence

$$\int_{S_{\alpha\beta}} \nabla^s\cdot\mathbf{f} dS = \nabla^S\cdot \int_{S_{\alpha\beta}} \mathbf{f}^s dS + \int_{S_{\alpha\beta}} (\nabla^s\cdot\mathbf{n})\,\mathbf{n}\cdot\mathbf{f}^c dS + \int_{C_{\alpha\beta\epsilon}} \boldsymbol{\nu}\cdot\mathbf{f}^s dC + \int_{C_{\alpha\beta_{\text{ends}}}} \frac{\mathbf{n}^*\cdot\mathbf{f}}{\boldsymbol{\nu}\cdot\mathbf{n}^*} dC \tag{8.30}$$

C[2, (2, 0), 1] Curl

$$\int_{S_{\alpha\beta}} \nabla^s \times \mathbf{f}^s dS = \nabla^S \times \int_{S_{\alpha\beta}} \mathbf{f} dS - \nabla^S\cdot \int_{S_{\alpha\beta}} \mathbf{nn} \times \mathbf{f}^s dS + \int_{S_{\alpha\beta}} (\nabla^s\cdot\mathbf{n})\,\mathbf{n} \times \mathbf{f}^s dS + \int_{C_{\alpha\beta\epsilon}} \boldsymbol{\nu} \times \mathbf{f} dC + \int_{C_{\alpha\beta_{\text{ends}}}} \boldsymbol{\nu} \times \mathbf{f} dC \tag{8.31}$$

T[2, (2, 0), 1] Transport

$$\int_{S_{\alpha\beta}} \left.\frac{\partial f}{\partial t}\right|_{\mathbf{u}} dS = \frac{\partial}{\partial t} \int_{S_{\alpha\beta}} f dS + \nabla^S\cdot \int_{S_{\alpha\beta}} \mathbf{w}^c f dS - \int_{S_{\alpha\beta}} (\nabla^s\cdot\mathbf{n})\,\mathbf{n}\cdot\mathbf{w}^c f dS - \int_{C_{\alpha\beta\epsilon}} \boldsymbol{\nu}\cdot\mathbf{w}^s f dC - \int_{C_{\alpha\beta_{\text{ends}}}} \frac{\mathbf{n}^*\cdot\mathbf{w}f}{\boldsymbol{\nu}\cdot\mathbf{n}^*} dC \tag{8.32}$$

8.3.3 Curvilineal Operator Theorems, [1, (2, 0), 1]

Figure and Notation

Refer to figure 8.6

$C_{\alpha\beta\epsilon}$ is the contact curve within the REV that is the location where the α- β- and ϵ-phases meet,

pts refers to the end points of $C_{\alpha\beta\epsilon}$ within the REV where the α- β- and ϵ-phases meet a fourth phase,

ends$_{\alpha\beta\epsilon}$ is the points at the ends of the REV formed by the intersection of the bounding surface of the cylinder with $C_{\alpha\beta\epsilon}$,

$\mathbf{n}^*$ is the unit vector normal to the external boundary of the REV pointing outward from the REV,

$\boldsymbol{\lambda}$ is the unit vector tangent to $C_{\alpha\beta\epsilon}$,

$\mathbf{e}$ is the unit vector tangent to $C_{\alpha\beta\epsilon}$ at pts and ends$_{\alpha\beta\epsilon}$ pointing outward from the curve such that $\mathbf{e}\cdot\boldsymbol{\lambda} = \pm 1$,

$\boldsymbol{\Lambda}$ is the unit vector tangent to the axis of the REV,

∇^c is the microscopic curvilineal operator, $\nabla^c = \boldsymbol{\lambda}\boldsymbol{\lambda}\cdot\nabla_\xi$,

∇^S is the two-dimensional macroscopic spatial operator in the directions orthogonal to the axis of the REV, $\nabla^S = \nabla_\mathbf{x} - \boldsymbol{\Lambda}\boldsymbol{\Lambda}\cdot\nabla_\mathbf{x}$,

$\mathbf{w}$ is the velocity of $C_{\alpha\beta\epsilon}$,

$\mathbf{w}^c$ is the velocity of $C_{\alpha\beta\epsilon}$ tangent to $C_{\alpha\beta\epsilon}$, $\mathbf{w}^c = \boldsymbol{\lambda}\boldsymbol{\lambda}\cdot\mathbf{w}$,

$\mathbf{w}^s$ is the velocity of $C_{\alpha\beta\epsilon}$ normal to $C_{\alpha\beta\epsilon}$, $\mathbf{w}^s = \mathbf{w} - \mathbf{w}^c$,

f is a curvilineal function $f(\mathbf{l}, t)$ defined on $C_{\alpha\beta\epsilon}$,

$\mathbf{f}$ is a curvilineal vector $\mathbf{f}(\mathbf{l}, t)$ defined on $C_{\alpha\beta\epsilon}$,

$\mathbf{f}^c$ is the vector tangent to $C_{\alpha\beta\epsilon}$ such that $\mathbf{f}^c = \boldsymbol{\lambda}\boldsymbol{\lambda}\cdot\mathbf{f}^c$, and

$\mathbf{f}^s$ is the vector normal to $C_{\alpha\beta\epsilon}$ such that $\mathbf{f}^s = \mathbf{f} - \mathbf{f}^c$.

G[1, (2, 0), 1] Gradient

$$\int_{C_{\alpha\beta\epsilon}} \nabla^c f dC = \nabla^S \cdot \int_{C_{\alpha\beta\epsilon}} \boldsymbol{\lambda}\boldsymbol{\lambda} f dC - \int_{C_{\alpha\beta\epsilon}} (\boldsymbol{\lambda}\cdot\nabla^c\boldsymbol{\lambda}) f dC + \sum_{\text{pts}} \mathbf{e} f + \sum_{\text{ends}_{\alpha\beta\epsilon}} \mathbf{e} f \quad \textbf{(8.33)}$$

D[1, (2, 0), 1] Divergence

$$\int_{C_{\alpha\beta\epsilon}} \nabla^c\cdot\mathbf{f} dC = \nabla^S \cdot \int_{C_{\alpha\beta\epsilon}} \mathbf{f}^c dC - \int_{C_{\alpha\beta\epsilon}} (\boldsymbol{\lambda}\cdot\nabla^c\boldsymbol{\lambda})\cdot\mathbf{f}^s dC + \sum_{\text{pts}} \mathbf{e}\cdot\mathbf{f}^c + \sum_{\text{ends}_{\alpha\beta\epsilon}} \frac{\mathbf{n}^*\cdot\mathbf{f}}{\mathbf{e}\cdot\mathbf{n}^*}$$

(8.34)

C[1, (2, 0), 1] Curl

$$\int_{C_{\alpha\beta\epsilon}} \nabla^c \times \mathbf{f} dC = \nabla^S \cdot \int_{C_{\alpha\beta\epsilon}} \boldsymbol{\lambda\lambda} \times \mathbf{f}^s dC - \int_{C_{\alpha\beta\epsilon}} (\boldsymbol{\lambda} \cdot \nabla^c \boldsymbol{\lambda}) \times \mathbf{f} dC$$

$$+ \sum_{\text{pts}} \mathbf{e} \times \mathbf{f}^s + \sum_{\text{ends}_{\alpha\beta\epsilon}} \frac{\mathbf{n}^* \times \mathbf{f}}{\mathbf{e} \cdot \mathbf{n}^*} \tag{8.35}$$

T[1, (2, 0), 1] Transport

$$\int_{C_{\alpha\beta\epsilon}} \left.\frac{\partial f}{\partial t}\right|_1 dC = \frac{\partial}{\partial t} \int_{C_{\alpha\beta\epsilon}} f dC + \nabla^S \cdot \int_{C_{\alpha\beta\epsilon}} \mathbf{w}^s f dC + \int_{C_{\alpha\beta\epsilon}} (\boldsymbol{\lambda} \cdot \nabla^c \boldsymbol{\lambda}) \cdot \mathbf{w}^s f dC$$

$$- \sum_{\text{pts}} \mathbf{e} \cdot \mathbf{w}^c f - \sum_{\text{ends}_{\alpha\beta\epsilon}} \frac{\mathbf{n}^* \cdot \mathbf{w} f}{\mathbf{e} \cdot \mathbf{n}^*} \tag{8.36}$$

8.4 TWO MACROSCOPIC, ONE MEGASCOPIC DIMENSION, [*i*, (0, 2), 1]

The averaging volume or REV for this set of averaging theorems is also the circular cylinder of figures 8.4 through 8.6. The radius of the cylinder is independent of position and time. The orientation of the axis of the cylinder is independent of time but may vary with position as long as this variation with position is such that the radius of curvature of the surface orthogonal to the axes of the REV's is much less than the diameter of the cylinder such that $|\nabla^S \cdot \boldsymbol{\Lambda}| D \ll 1$. Changes in the boundary of the REV only occur at the intersection of the REV with the boundary such that the effective length and orientation of ends of the REV's is not constant. Theorems in this section will not contain macroscopic derivatives but are formulated in terms of fluxes across the boundary of the REV. This change in the theorems from those found in Section 8.3 is indicated by the interchange of the second and third indices of the theorem family designation.

8.4.1 Spatial Operator Theorems, [3, (0, 2), 1]

Figure and Notation

Refer to figure 8.4

V_α is the portion of the REV occupied by the α-phase,

S_α is the portion of the external boundary of the REV that intersects the α-phase, including the end boundaries $S_{\alpha_{\text{ends}}}$,

$S_{\alpha\beta}$ is the surface within the REV between the α-phase and all other phases,

$S_{\alpha_{\text{ends}}}$ is a subset of S_α composed of the portion of the end boundaries of the REV occupied by the α-phase,

$\mathbf{n}^*$ is the unit vector normal to the external boundary of the REV pointing outward from the REV,

$\mathbf{n}$ is the unit vector normal to the $S_{\alpha\beta}$ surface pointing outward from the α-phase,

∇ is the microscopic spatial operator, $\nabla = \nabla_\xi$,

$\mathbf{w}$ is the velocity of either $S_{\alpha\beta}$ or $S_{\alpha_{\text{ends}}}$,

$\mathbf{w}^c$ is the velocity of the surface $S_{\alpha\beta}$ normal to $S_{\alpha\beta}$, $\mathbf{w}^c = \mathbf{nn}\bullet\mathbf{w}$

f is a spatial function $f(\mathbf{x}, t)$,

$\mathbf{f}$ is a spatial vector $\mathbf{f}(\mathbf{x}, t)$,

$\mathbf{f}^c$ is the vector normal to $S_{\alpha\beta}$ such that $\mathbf{f}^c = \mathbf{nn}\bullet\mathbf{f}$, and

$\mathbf{f}^s$ is the vector tangent to $S_{\alpha\beta}$ such that $\mathbf{f}^s = \mathbf{f} - \mathbf{f}^c$.

G[3, (0, 2), 1] Gradient

$$\int_{V_\alpha} \nabla f dV = \int_{S_\alpha} \mathbf{n}^* f dS + \int_{S_{\alpha\beta}} \mathbf{n} f dS \tag{8.37}$$

D[3, (0, 2), 1] Divergence

$$\int_{V_\alpha} \nabla\bullet\mathbf{f} dV = \int_{S_\alpha} \mathbf{n}^*\bullet\mathbf{f} dS + \int_{S_{\alpha\beta}} \mathbf{n}\bullet\mathbf{f}^c dS \tag{8.38}$$

C[3, (0, 2), 1] Curl

$$\int_{V_\alpha} \nabla\times\mathbf{f} dV = \int_{S_\alpha} \mathbf{n}^*\times\mathbf{f} dS + \int_{S_{\alpha\beta}} \mathbf{n}\times\mathbf{f}^s dS \tag{8.39}$$

T[3, (0, 2), 1] Transport

$$\int_{V_\alpha} \left.\frac{\partial f}{\partial t}\right|_{\mathbf{x}} dV = \frac{\partial}{\partial t}\int_{V_\alpha} f dV - \int_{S_{\alpha_{\text{ends}}}} \mathbf{n}^*\bullet\mathbf{w} f dS - \int_{S_{\alpha\beta}} \mathbf{n}\bullet\mathbf{w}^c f dS \tag{8.40}$$

8.4.2 Surficial Operator Theorems, [2, (0, 2), 1]

Figure and Notation

Refer to figure 8.5

$S_{\alpha\beta}$ is the surface within the REV between the α- and the β-phases,

$C_{\alpha\beta}$ is the curve of intersection of $S_{\alpha\beta}$ with the external surface of the REV,

$C_{\alpha\beta\epsilon}$ is the boundary curve of $S_{\alpha\beta}$ within the REV that also is the location where the α- and β-phases meet a third phase,

$C_{\alpha\beta_{\text{sides}}}$ is the part of $C_{\alpha\beta}$ at the sides of the cylindrical REV formed by the intersection of $S_{\alpha\beta}$ with the sides of the cylinder,

$\mathbf{n}^*$ is the unit vector normal to the external boundary of the REV pointing outward from the REV,

$\mathbf{n}$ is the unit vector normal to the $S_{\alpha\beta}$ surface,

$\boldsymbol{\nu}$ is the unit vector tangent to the $S_{\alpha\beta}$ surface and normal to $C_{\alpha\beta\epsilon}$ and $C_{\alpha\beta}$ such that $\mathbf{n}\bullet\boldsymbol{\nu} = 0$,

∇^s is the microscopic surficial operator, $\nabla^s = \nabla_\xi - \mathbf{nn}\bullet\nabla_\xi$,

$\mathbf{w}$ is the velocity of $S_{\alpha\beta}$,

$\mathbf{w}^c$ is the velocity of $S_{\alpha\beta}$ normal to $S_{\alpha\beta}$, $\mathbf{w}^c = \mathbf{nn}\bullet\mathbf{w}$

$\mathbf{w}^s$ is the velocity of $S_{\alpha\beta}$ tangent to $S_{\alpha\beta}$, $\mathbf{w}^s = \mathbf{w} - \mathbf{w}^c$,

f is a surficial function $f(\mathbf{u}, t)$ defined on $S_{\alpha\beta}$,

$\mathbf{f}$ is a surficial vector $\mathbf{f}(\mathbf{u}, t)$ defined on $S_{\alpha\beta}$,

$\mathbf{f}^c$ is the vector normal to $S_{\alpha\beta}$ such that $\mathbf{f}^c = \mathbf{nn}\bullet\mathbf{f}$, and

$\mathbf{f}^s$ is the vector tangent to $S_{\alpha\beta}$ such that $\mathbf{f}^s = \mathbf{f} - \mathbf{f}^c$.

G[2, (0, 2), 1] Gradient

$$\int_{S_{\alpha\beta}} \nabla^s f dS = \int_{S_{\alpha\beta}} (\nabla^s\bullet\mathbf{n})\,\mathbf{n} f dS + \int_{C_{\alpha\beta}} \boldsymbol{\nu} f dS + \int_{C_{\alpha\beta\epsilon}} \boldsymbol{\nu} f dS \qquad \textbf{(8.41)}$$

D[2, (0, 2), 1] Divergence

$$\int_{S_{\alpha\beta}} \nabla^s\bullet\mathbf{f} dS = \int_{S_{\alpha\beta}} (\nabla^s\bullet\mathbf{n})\,\mathbf{n}\bullet\mathbf{f}^c dS + \int_{C_{\alpha\beta}} \boldsymbol{\nu}\bullet\mathbf{f}^s dC + \int_{C_{\alpha\beta\epsilon}} \boldsymbol{\nu}\bullet\mathbf{f}^s dC \qquad \textbf{(8.42)}$$

C[2, (0, 2), 1] Curl

$$\int_{S_{\alpha\beta}} \nabla^s \times \mathbf{f} dS = \int_{S_{\alpha\beta}} (\nabla^s \bullet \mathbf{n})\, \mathbf{n} \times \mathbf{f}^s dS + \int_{C_{\alpha\beta}} \boldsymbol{\nu} \times \mathbf{f}^s dC + \int_{C_{\alpha\beta\epsilon}} \boldsymbol{\nu} \times \mathbf{f}^s dC \quad \textbf{(8.43)}$$

T[2, (0, 2), 1] Transport

$$\int_{S_{\alpha\beta}} \left.\frac{\partial f}{\partial t}\right|_{\mathbf{u}} dS = \frac{\partial}{\partial t} \int_{S_{\alpha\beta}} f dS - \int_{S_{\alpha\beta}} (\nabla^s \bullet \mathbf{n})\, \mathbf{n} \bullet \mathbf{w}^c f dS + \int_{C_{\alpha\beta_{\text{sides}}}} \frac{\mathbf{n}^* \bullet \mathbf{w} f}{\boldsymbol{\nu} \bullet \mathbf{n}^*} dC$$

$$- \int_{C_{\alpha\beta}} \boldsymbol{\nu} \bullet \mathbf{w}^s f dC - \int_{C_{\alpha\beta\epsilon}} \boldsymbol{\nu} \bullet \mathbf{w}^s f dC \quad \textbf{(8.44)}$$

8.4.3 Curvilineal Operator Theorems, [1, (0, 2), 1]

Figure and Notation

Refer to figure 8.6

- $C_{\alpha\beta\epsilon}$ is the contact curve within the REV that exists as the location where the α- β- and ϵ-phases meet,
- ext refers to the points of intersection of the $C_{\alpha\beta\epsilon}$ curve with the external boundary of the REV,
- pts refers to the end points of the $C_{\alpha\beta\epsilon}$ curve within the REV where the α- β- and ϵ-phases meet a fourth phase,
- $\text{sides}_{\alpha\beta\epsilon}$ is the subset of the points comprising ext that are on the sides of the cylindrical REV formed by the intersection of bounding surface of the cylinder with $C_{\alpha\beta\epsilon}$,
- $\mathbf{n}^*$ is the unit vector normal to the external boundary of the REV pointing outward from the REV,
- $\boldsymbol{\lambda}$ is the unit vector tangent to $C_{\alpha\beta\epsilon}$,
- $\mathbf{e}$ is the unit vector tangent to $C_{\alpha\beta\epsilon}$ at pts and ext pointing outward from the curve such that $\mathbf{e} \bullet \boldsymbol{\lambda} = \pm 1$,
- ∇^c is the microscopic curvilineal operator, $\nabla^c = \boldsymbol{\lambda\lambda} \bullet \nabla_{\xi}$,
- $\mathbf{w}$ is the velocity of $C_{\alpha\beta\epsilon}$,
- $\mathbf{w}^c$ is the velocity of $C_{\alpha\beta\epsilon}$ tangent to $C_{\alpha\beta\epsilon}$, $\mathbf{w}^c = \boldsymbol{\lambda\lambda} \bullet \mathbf{w}$,
- $\mathbf{w}^s$ is the velocity of $C_{\alpha\beta\epsilon}$ normal to $C_{\alpha\beta\epsilon}$, $\mathbf{w}^s = \mathbf{w} - \mathbf{w}^c$,
- f is a curvilineal function $f(\mathbf{l}, t)$ defined on $C_{\alpha\beta\epsilon}$,
- $\mathbf{f}$ is a curvilineal vector $\mathbf{f}(\mathbf{l}, t)$ defined on $C_{\alpha\beta\epsilon}$,
- $\mathbf{f}^c$ is the vector tangent to $C_{\alpha\beta\epsilon}$ such that $\mathbf{f}^c = \boldsymbol{\lambda\lambda} \bullet \mathbf{f}$, and
- $\mathbf{f}^s$ is the vector normal to $C_{\alpha\beta\epsilon}$ such that $\mathbf{f}^s = \mathbf{f} - \mathbf{f}^c$.

G[1, (0, 2), 1] Gradient

$$\int_{C_{\alpha\beta\epsilon}} \nabla^c f dC = -\int_{C_{\alpha\beta\epsilon}} (\boldsymbol{\lambda}\cdot\nabla^c\boldsymbol{\lambda}) f dC + \sum_{\text{pts}} \mathbf{e}f + \sum_{\text{ext}} \mathbf{e}f \tag{8.45}$$

D[1, (0, 2), 1] Divergence

$$\int_{C_{\alpha\beta\epsilon}} \nabla^c\cdot\mathbf{f}^c dC = -\int_{C_{\alpha\beta\epsilon}} (\boldsymbol{\lambda}\cdot\nabla^c\boldsymbol{\lambda})\cdot\mathbf{f}^s dC + \sum_{\text{pts}} \mathbf{e}\cdot\mathbf{f}^c + \sum_{\text{ext}} \mathbf{e}\cdot\mathbf{f}^c \tag{8.46}$$

C[1, (0, 2), 1] Curl

$$\int_{C_{\alpha\beta\epsilon}} \nabla^c\times\mathbf{f} dC = -\int_{C_{\alpha\beta\epsilon}} (\boldsymbol{\lambda}\cdot\nabla^c\boldsymbol{\lambda})\times\mathbf{f} dC + \sum_{\text{pts}} \mathbf{e}\times\mathbf{f}^s + \sum_{\text{ext}} \mathbf{e}\times\mathbf{f}^s \tag{8.47}$$

T[1, (0, 2), 1] Transport

$$\int_{C_{\alpha\beta\epsilon}} \left.\frac{\partial f}{\partial t}\right|_1 dC = \frac{\partial}{\partial t}\int_{C_{\alpha\beta\epsilon}} f dC + \int_{C_{\alpha\beta\epsilon}} \boldsymbol{\lambda}\cdot\nabla^c\boldsymbol{\lambda}\cdot\mathbf{w}^s f dC - \sum_{\text{pts}} \mathbf{e}\cdot\mathbf{w}^c f$$

$$+ \sum_{\text{sides}_{\alpha\beta\epsilon}} \frac{\mathbf{n}^*\cdot\mathbf{w}f}{\mathbf{e}\cdot\mathbf{n}^*} - \sum_{\text{ext}} \mathbf{e}\cdot\mathbf{w}^c f \tag{8.48}$$

8.5 ONE MACROSCOPIC, TWO MEGASCOPIC DIMENSIONS, [*i*, (1, 0), 2]

The averaging volume or REV for this set of theorems is a slab of finite thickness, D, with parallel faces as depicted in figures 8.7 through 8.9. The faces of the REV intersect the megascopic boundary of the system of interest along their edges. The macroscopic direction is perpendicular to the faces of the slab, and the two megascopic directions are tangent to the slab faces. The thickness of the slab is constant in time and space but the orientation of this REV may vary with position as long as $|\mathbf{N}\cdot\nabla_{\mathbf{x}}\mathbf{N}|D \ll 1$. The volume of the effective REV may not be constant due to changes in the intersection of the slab with the physical megascopic boundaries of a system that does not have a constant cross-section or whose cross-section changes with time. The theorems presented in this section will contain macroscopic spatial derivatives in the direction perpendicular to the slab face and flux terms along the edges of the slab.

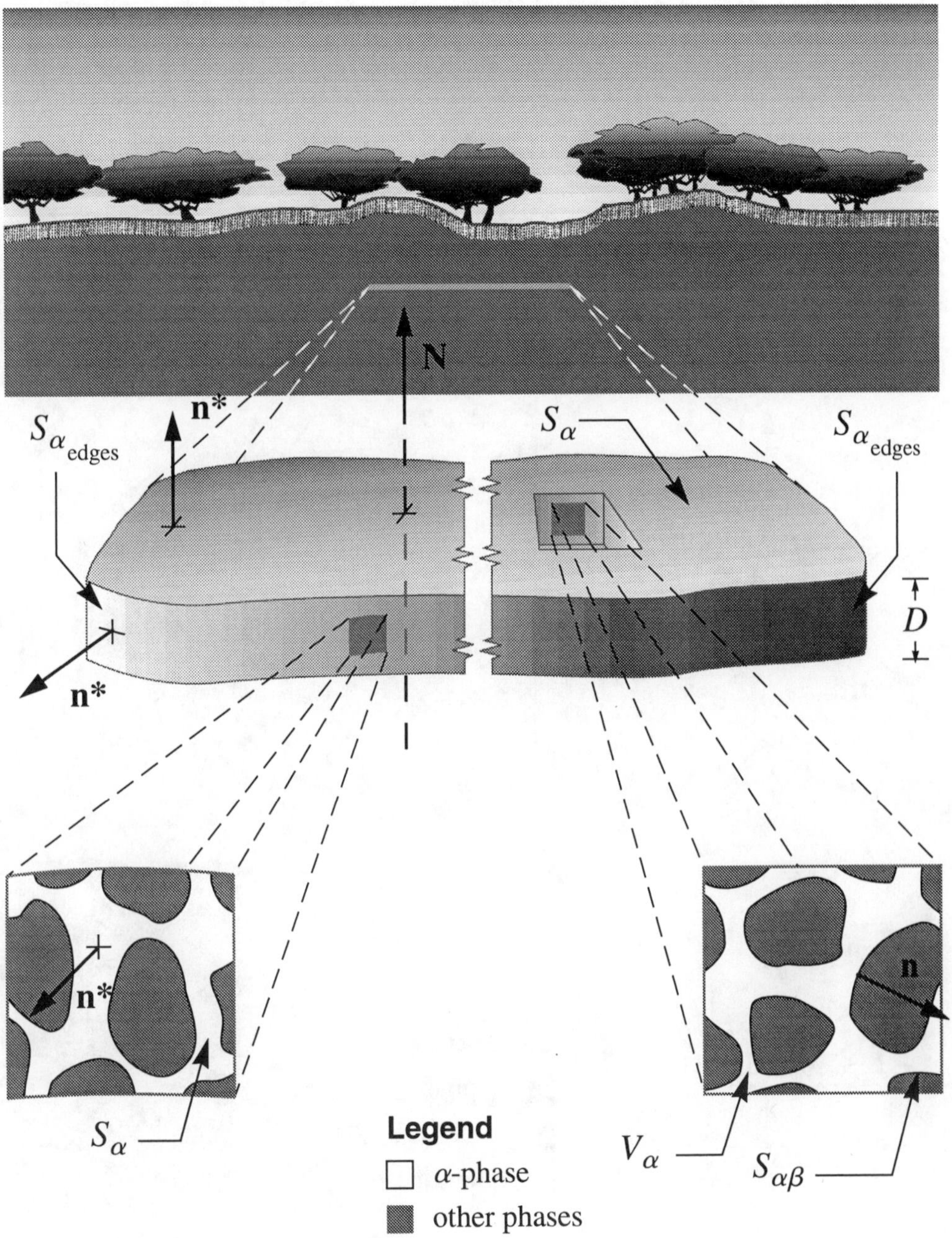

Figure 8.7. Slab REV with constant thickness but extent and orientation dependent on space and time (wedge removed for illustrative purposes), used to develop averaging theorems for spatial operators, equations (8.49) through (8.52) and (8.61) through (8.64).

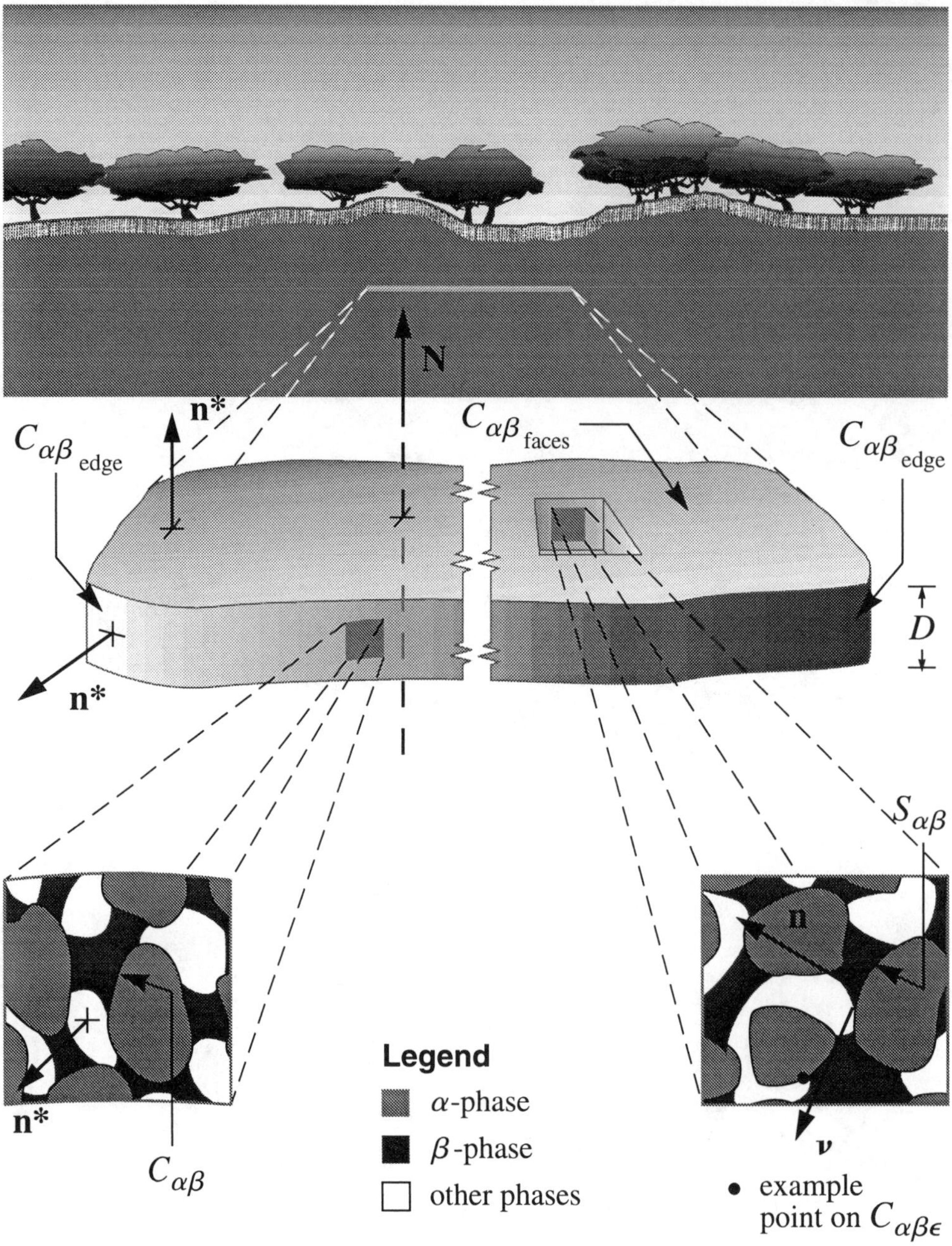

Figure 8.8. Slab REV with constant thickness but extent and orientation dependent on space and time (wedge removed for illustrative purposes), used to develop averaging theorems for surficial operators, equations (8.53) through (8.56) and (8.65) through (8.68).

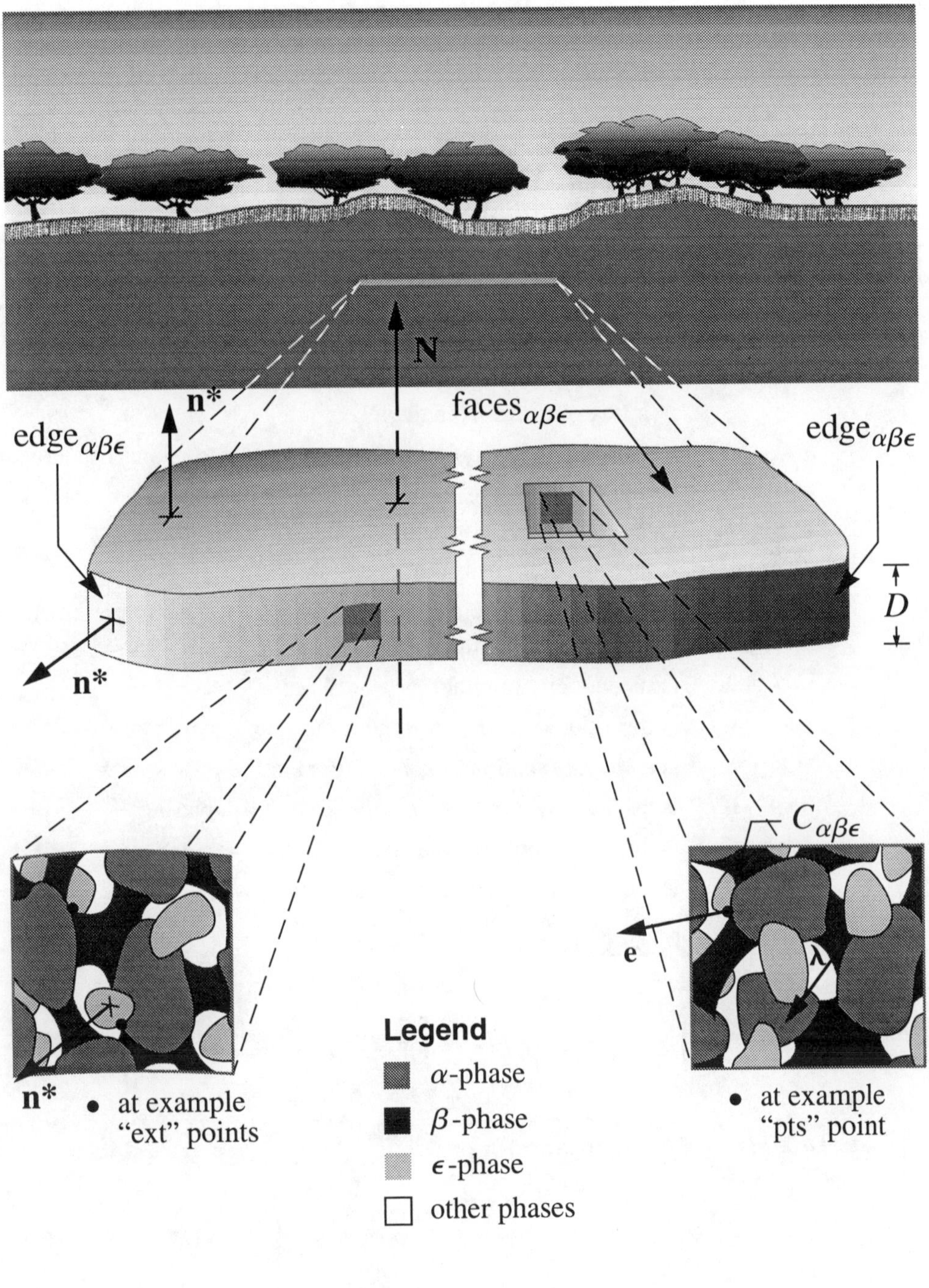

Figure 8.9. Slab REV with constant thickness but extent and orientation dependent on space and time (wedge removed for illustrative purposes), used to develop averaging theorems for curvilineal operators, equations (8.57) through (8.60) and (8.69) through (8.72).

8.5.1 Spatial Operator Theorems, [3, (1, 0), 2]

Figure and Notation

Refer to figure 8.7

- V_α is the portion of the REV occupied by the α-phase,
- $S_{\alpha\beta}$ is the surface within the REV between the α-phase and all other phases,
- $S_{\alpha_{\text{edges}}}$ is the portion of the edge boundaries of the REV that is occupied by the α-phase,
- $\mathbf{n}^*$ is the unit vector normal to the external boundary of the REV pointing outward from the REV,
- $\mathbf{n}$ is the unit vector normal on the $S_{\alpha\beta}$ surface pointing outward from the α-phase,
- $\mathbf{N}$ is the unit vector normal to the face of the slab REV,
- ∇ is the microscopic spatial operator, $\nabla = \nabla_\xi$,
- ∇^C is the one-dimensional macroscopic spatial operator in the direction normal to the face of the slab REV, $\nabla^C = \mathbf{NN}\bullet\nabla_\mathbf{x}$,
- $\mathbf{w}$ is the velocity of either $S_{\alpha\beta}$ or $S_{\alpha_{\text{edges}}}$,
- $\mathbf{w}^c$ is the velocity of $S_{\alpha\beta}$ normal to $S_{\alpha\beta}$, $\mathbf{w}^c = \mathbf{nn}\bullet\mathbf{w}$,
- f is a spatial function $f(\mathbf{x}, t)$ defined in the α-phase,
- $\mathbf{f}$ is a spatial vector $\mathbf{f}(\mathbf{x}, t)$ defined in the α-phase,
- $\mathbf{f}^c$ is the vector normal to $S_{\alpha\beta}$ such that $\mathbf{f}^c = \mathbf{nn}\bullet\mathbf{f}$, and
- $\mathbf{f}^s$ is the vector tangent to $S_{\alpha\beta}$ such that $\mathbf{f}^s = \mathbf{f} - \mathbf{f}^c$.

G[3, (1, 0), 2] Gradient

$$\int_{V_\alpha} \nabla f dV = \nabla^C \int_{V_\alpha} f dV + \int_{S_{\alpha\beta}} \mathbf{n} f dS + \int_{S_{\alpha_{\text{edges}}}} \mathbf{n}^* f dS \qquad (8.49)$$

D[3, (1, 0), 2] Divergence

$$\int_{V_\alpha} \nabla\bullet\mathbf{f} dV = \nabla^C\bullet \int_{V_\alpha} \mathbf{f} dV + \int_{S_{\alpha\beta}} \mathbf{n}\bullet\mathbf{f}^c dS + \int_{S_{\alpha_{\text{edges}}}} \mathbf{n}^*\bullet\mathbf{f} dS \qquad (8.50)$$

C[3, (1, 0), 2] Curl

$$\int_{V_\alpha} \nabla\times\mathbf{f} dV = \nabla^C\times \int_{V_\alpha} \mathbf{f} dV + \int_{S_{\alpha\beta}} \mathbf{n}\times\mathbf{f}^s dS + \int_{S_{\alpha_{\text{edges}}}} \mathbf{n}^*\times\mathbf{f} dS \qquad (8.51)$$

T[3, (1, 0), 2] Transport

$$\int_{V_\alpha} \left.\frac{\partial f}{\partial t}\right|_{\mathbf{x}} dV = \frac{\partial}{\partial t}\int_{V_\alpha} f dV - \int_{S_{\alpha\beta}} \mathbf{n}\bullet\mathbf{w}^c f dS - \int_{S_{\alpha_{\text{edges}}}} \mathbf{n}^*\bullet\mathbf{w} f dS \qquad \textbf{(8.52)}$$

8.5.2 Surficial Operator Theorems, [2, (1, 0), 2]

Figure and Notation

Refer to figure 8.8

$S_{\alpha\beta}$ is the surface within the slab REV between the α- and the β-phases,

$C_{\alpha\beta\epsilon}$ is the boundary curve of $S_{\alpha\beta}$ within the REV that also is the location where the α- and β-phases meet a third phase,

$C_{\alpha\beta_{\text{edge}}}$ is the curve at the edge of the REV formed by the intersection of $S_{\alpha\beta}$ with the edge of the slab,

$\mathbf{n}^*$ is the unit vector normal to the external boundary of the REV pointing outward from the REV,

$\mathbf{n}$ is the unit vector normal to the $S_{\alpha\beta}$ surface,

$\boldsymbol{\nu}$ is the unit vector tangent to the $S_{\alpha\beta}$ surface and normal to $C_{\alpha\beta\epsilon}$ and $C_{\alpha\beta_{\text{edge}}}$ such that $\mathbf{n}\bullet\boldsymbol{\nu} = 0$,

$\mathbf{N}$ is the unit vector normal to the face of the slab REV,

∇ is the microscopic spatial operator, $\nabla = \nabla_\xi$,

∇^s is the microscopic surficial operator, $\nabla^s = \nabla_\xi - \mathbf{nn}\bullet\nabla_\xi$,

∇^C is the one-dimensional macroscopic spatial operator in the direction normal to the face of the slab REV, $\nabla^C = \mathbf{NN}\bullet\nabla_\mathbf{x}$,

$\mathbf{w}$ is the velocity of either $S_{\alpha\beta}$ or $S_{\alpha_{\text{edges}}}$,

$\mathbf{w}^c$ is the velocity of $S_{\alpha\beta}$ normal to $S_{\alpha\beta}$ such that $\mathbf{w}^c = \mathbf{nn}\bullet\mathbf{w}$,

$\mathbf{w}^s$ is the velocity of $S_{\alpha\beta}$ tangent to $S_{\alpha\beta}$ such that $\mathbf{w}^s = \mathbf{w} - \mathbf{w}^c$,

f is a surficial function $f(\mathbf{u}, t)$ defined on $S_{\alpha\beta}$,

$\mathbf{f}$ is a surficial vector $\mathbf{f}(\mathbf{u}, t)$ defined on $S_{\alpha\beta}$,

$\mathbf{f}^c$ is the vector normal to $S_{\alpha\beta}$ such that $\mathbf{f}^c = \mathbf{nn}\bullet\mathbf{f}$, and

$\mathbf{f}^s$ is the vector tangent to $S_{\alpha\beta}$ such that $\mathbf{f}^s = \mathbf{f} - \mathbf{f}^c$.

G[2, (1, 0), 2] Gradient

$$\int_{S_{\alpha\beta}} \nabla^s f dS = \nabla^C \int_{S_{\alpha\beta}} f dS - \nabla^C\bullet \int_{S_{\alpha\beta}} \mathbf{nn} f dS + \int_{S_{\alpha\beta}} (\nabla^s\bullet\mathbf{n})\,\mathbf{n} f dS + \int_{C_{\alpha\beta\epsilon}} \boldsymbol{\nu} f dC + \int_{C_{\alpha\beta_{\text{edge}}}} \boldsymbol{\nu} f dC$$

(8.53)

D[2, (1, 0), 2] Divergence

$$\int_{S_{\alpha\beta}} \nabla^s \cdot \mathbf{f} dS = \nabla^C \cdot \int_{S_{\alpha\beta}} \mathbf{f}^s dS + \int_{S_{\alpha\beta}} (\nabla^s \cdot \mathbf{n})\, \mathbf{n} \cdot \mathbf{f}^c dS + \int_{C_{\alpha\beta\epsilon}} \boldsymbol{\nu} \cdot \mathbf{f}^s dC + \int_{C_{\alpha\beta_{\text{edge}}}} \boldsymbol{\nu} \cdot \mathbf{f}^s dC \tag{8.54}$$

C[2, (1, 0), 2] Curl

$$\int_{S_{\alpha\beta}} \nabla^s \times \mathbf{f} dS = \nabla^C \times \int_{S_{\alpha\beta}} \mathbf{f} dS - \nabla^C \cdot \int_{S_{\alpha\beta}} \mathbf{n}\mathbf{n} \times \mathbf{f}^s dS + \int_{S_{\alpha\beta}} (\nabla^s \cdot \mathbf{n})\, \mathbf{n} \times \mathbf{f}^s dS$$

$$+ \int_{C_{\alpha\beta\epsilon}} \boldsymbol{\nu} \times \mathbf{f} dC + \int_{C_{\alpha\beta_{\text{edge}}}} \boldsymbol{\nu} \times \mathbf{f} dC \tag{8.55}$$

T[2, (1, 0), 2] Transport

$$\int_{S_{\alpha\beta}} \left.\frac{\partial f}{\partial t}\right|_{\mathbf{u}} dS = \frac{\partial}{\partial t} \int_{S_{\alpha\beta}} f dS + \nabla^C \cdot \int_{S_{\alpha\beta}} \mathbf{w}^c f dS - \int_{S_{\alpha\beta}} (\nabla^s \cdot \mathbf{n})\, \mathbf{n} \cdot \mathbf{w}^c f dS$$

$$- \int_{C_{\alpha\beta\epsilon}} \boldsymbol{\nu} \cdot \mathbf{w}^s f dC - \int_{C_{\alpha\beta_{\text{edge}}}} \boldsymbol{\nu} \cdot \mathbf{w}^s f dC \tag{8.56}$$

8.5.3 Curvilineal Operator Theorems, [1, (1, 0), 2]

Figure and Notation

Refer to figure 8.9

$C_{\alpha\beta\epsilon}$ is the contact curve within the slab REV that is the location where the α- β- and ϵ-phases meet,

pts refers to the end points of the $C_{\alpha\beta\epsilon}$ curve within the REV where the α- β- and ϵ-phases meet a fourth phase,

$\text{edge}_{\alpha\beta\epsilon}$ is the points at the edge of the REV formed by the intersection of bounding surface of the slab with $C_{\alpha\beta\epsilon}$,

$\mathbf{n}^*$ is the unit vector normal to the external boundary of the REV pointing outward from the REV,

$\boldsymbol{\lambda}$ is the unit vector tangent to $C_{\alpha\beta\epsilon}$,

$\mathbf{e}$ is the unit vector tangent to $C_{\alpha\beta\epsilon}$ at pts and $\text{edge}_{\alpha\beta\epsilon}$ pointing outward from the curve such that $\mathbf{e} \cdot \boldsymbol{\lambda} = \pm 1$,

N is the unit vector normal to the face of the slab REV,
∇^c is the microscopic curvilineal operator, $\nabla^c = \boldsymbol{\lambda\lambda}\bullet\nabla_\xi$,
∇^C is the one-dimensional macroscopic spatial operator in the direction normal to the face of the slab REV, $\nabla^C = \mathbf{NN}\bullet\nabla_\mathbf{x}$,
w is the velocity of $C_{\alpha\beta\epsilon}$,
$\mathbf{w}^c$ is the velocity of $C_{\alpha\beta\epsilon}$ tangent to $C_{\alpha\beta\epsilon}$, $\mathbf{w}^c = \boldsymbol{\lambda\lambda}\bullet\mathbf{w}$,
$\mathbf{w}^s$ is the velocity of $C_{\alpha\beta\epsilon}$ normal to $C_{\alpha\beta\epsilon}$, $\mathbf{w}^s = \mathbf{w} - \mathbf{w}^c$,
f is a curvilineal function $f(\mathbf{l}, t)$ defined on $C_{\alpha\beta\epsilon}$,
f is a curvilineal vector $\mathbf{f}(\mathbf{l}, t)$ defined on $C_{\alpha\beta\epsilon}$,
$\mathbf{f}^c$ is the vector tangent to $C_{\alpha\beta\epsilon}$ such that $\mathbf{f}^c = \boldsymbol{\lambda\lambda}\bullet\mathbf{f}$, and
$\mathbf{f}^s$ is the vector normal to $C_{\alpha\beta\epsilon}$ such that $\mathbf{f}^s = \mathbf{f} - \mathbf{f}^c$.

G[1, (1, 0), 2] Gradient

$$\int_{C_{\alpha\beta\epsilon}} \nabla^c f dC = \nabla^C \bullet \int_{C_{\alpha\beta\epsilon}} \boldsymbol{\lambda\lambda} f dC - \int_{C_{\alpha\beta\epsilon}} (\boldsymbol{\lambda}\bullet\nabla^c\boldsymbol{\lambda}) f dC + \sum_{\text{pts}} \mathbf{e} f + \sum_{\text{edge}_{\alpha\beta\epsilon}} \mathbf{e} f \tag{8.57}$$

D[1, (1, 0), 2] Divergence

$$\int_{C_{\alpha\beta\epsilon}} \nabla^c \bullet \mathbf{f} dC = \nabla^C \bullet \int_{C_{\alpha\beta\epsilon}} \mathbf{f}^c dC - \int_{C_{\alpha\beta\epsilon}} (\boldsymbol{\lambda}\bullet\nabla^c\boldsymbol{\lambda}) \bullet \mathbf{f}^s dC + \sum_{\text{pts}} \mathbf{e}\bullet\mathbf{f}^c + \sum_{\text{edge}_{\alpha\beta\epsilon}} \mathbf{e}\bullet\mathbf{f}^c \tag{8.58}$$

C[1, (1, 0), 2] Curl

$$\int_{C_{\alpha\beta\epsilon}} \nabla^c \times \mathbf{f} dC = \nabla^C \bullet \int_{C_{\alpha\beta\epsilon}} \boldsymbol{\lambda\lambda} \times \mathbf{f}^s dC - \int_{C_{\alpha\beta\epsilon}} (\boldsymbol{\lambda}\bullet\nabla^c\boldsymbol{\lambda}) \times \mathbf{f} dC + \sum_{\text{pts}} \mathbf{e}\times\mathbf{f}^s + \sum_{\text{edge}_{\alpha\beta\epsilon}} \mathbf{e}\times\mathbf{f}^s \tag{8.59}$$

T[1, (1, 0), 2] Transport

$$\int_{C_{\alpha\beta\epsilon}} \left.\frac{\partial f}{\partial t}\right|_\mathbf{l} dC = \frac{\partial}{\partial t} \int_{C_{\alpha\beta\epsilon}} f dC + \nabla^C \bullet \int_{C_{\alpha\beta\epsilon}} \mathbf{w}^s f dC + \int_{C_{\alpha\beta\epsilon}} \boldsymbol{\lambda}\bullet\nabla^c\mathbf{w}^s\bullet\boldsymbol{\lambda} dC + \int_{C_{\alpha\beta\epsilon}} \boldsymbol{\lambda}\bullet\nabla^c\boldsymbol{\lambda}\bullet\mathbf{w}^s f dC - \sum_{\text{pts}} \mathbf{e}\bullet\mathbf{w}^c f - \sum_{\text{edge}_{\alpha\beta\epsilon}} \mathbf{e}\bullet\mathbf{w}^c f \tag{8.60}$$

8.6 ONE MACROSCOPIC, TWO MEGASCOPIC DIMENSIONS, [*i*, (0, 1), 2]

The averaging volume for this set of theorems is the slab with parallel faces depicted in figures 8.7 through 8.9. The macroscopic direction is the direction perpendicular to the faces of the slab, and the megascopic directions are the directions tangent to the slab faces. The thickness of the slab is constant but the orientation of the slab may vary with position as long as $|\mathbf{N} \cdot \nabla_\mathbf{x} \mathbf{N}| D \ll 1$. The volume of the effective REV is not constant due to changes in the intersection of the slab with the physical megascopic boundaries of a system that does not have a constant cross-section. Theorems presented in this section are given in terms of fluxes on the REV boundary. No macroscopic spatial operator is retained.

8.6.1 Spatial Operator Theorems, [3, (0, 1), 2]

Figure and Notation

Refer to figure 8.7

V_α is the portion of the REV occupied by the α-phase,

S_α is the portion of the external boundary of the REV that intersects the α-phase, including the portion of the edge surface S_{α_edges},

$S_{\alpha\beta}$ is the surface within the REV between the α-phase and all other phases,

S_{α_edges} is the portion of the edge boundaries of the REV that is occupied by the α-phase,

$\mathbf{n}^*$ is the unit vector normal to the external boundary of the REV pointing outward from the REV,

$\mathbf{n}$ is the unit vector normal on the $S_{\alpha\beta}$ surface pointing outward from the α-phase,

∇ is the microscopic spatial operator, $\nabla = \nabla_\xi$,

$\mathbf{w}$ is the velocity of either $S_{\alpha\beta}$ or S_{α_edges},

$\mathbf{w}^c$ is the velocity of $S_{\alpha\beta}$ normal to $S_{\alpha\beta}$, $\mathbf{w}^c = \mathbf{n}\mathbf{n} \cdot \mathbf{w}$,

f is a spatial function $f(\mathbf{x}, t)$ defined in the α-phase,

$\mathbf{f}$ is a spatial vector $\mathbf{f}(\mathbf{x}, t)$ defined in the α-phase,

$\mathbf{f}^c$ is the vector normal to $S_{\alpha\beta}$ such that $\mathbf{f}^c = \mathbf{n}\mathbf{n} \cdot \mathbf{f}$, and

$\mathbf{f}^s$ is the vector tangent to $S_{\alpha\beta}$ such that $\mathbf{f}^s = \mathbf{f} - \mathbf{f}^c$.

G[3, (0, 1), 2] Gradient

$$\int_{V_\alpha} \nabla f dV = \int_{S_\alpha} \mathbf{n}^* f dS + \int_{S_{\alpha\beta}} \mathbf{n} f dS \tag{8.61}$$

D[3, (0, 1), 2] Divergence

$$\int_{V_\alpha} \nabla \cdot \mathbf{f} dV = \int_{S_\alpha} \mathbf{n}^* \cdot \mathbf{f} dS + \int_{S_{\alpha\beta}} \mathbf{n} \cdot \mathbf{f}^c dS \tag{8.62}$$

C[3, (0, 1), 2] Curl

$$\int_{V_\alpha} \nabla \times \mathbf{f} dV = \int_{S_\alpha} \mathbf{n}^* \times \mathbf{f} dS + \int_{S_{\alpha\beta}} \mathbf{n} \times \mathbf{f}^s dS \tag{8.63}$$

T[3, (0, 1), 2] Transport

$$\int_{V_\alpha} \left.\frac{\partial f}{\partial t}\right|_{\mathbf{x}} dV = \frac{\partial}{\partial t} \int_{V_\alpha} f dV - \int_{S_{\alpha_{\text{edges}}}} \mathbf{n}^* \cdot \mathbf{w} f dS - \int_{S_{\alpha\beta}} \mathbf{n} \cdot \mathbf{w}^c f dS \tag{8.64}$$

8.6.2 Surficial Operator Theorems, [2, (0, 1), 2]

Figure and Notation

Refer to figure 8.8

$S_{\alpha\beta}$ is the surface within the REV between the α- and the β-phases,

$C_{\alpha\beta}$ is the curve of intersection of $S_{\alpha\beta}$ with the external surface of the REV,

$C_{\alpha\beta\epsilon}$ is the boundary curve of $S_{\alpha\beta}$ within the REV that also is the location where the α- and β-phases meet a third phase,

$C_{\alpha\beta_{\text{faces}}}$ is the part of $C_{\alpha\beta}$ on the faces of the slab REV formed by the intersection of $S_{\alpha\beta}$ with the edge of the slab,

$\mathbf{n}^*$ is the unit vector normal to the external boundary of the REV pointing outward from the REV,

$\mathbf{n}$ is the unit vector normal to the $S_{\alpha\beta}$ surface,

$\boldsymbol{\nu}$ is the unit vector tangent to the $S_{\alpha\beta}$ surface and normal to $C_{\alpha\beta\epsilon}$ and $C_{\alpha\beta_{\text{faces}}}$ such that $\mathbf{n} \cdot \boldsymbol{\nu} = 0$,

$\mathbf{N}$ is the unit vector normal to the face of the slab REV,

∇ is the microscopic spatial operator, $\nabla = \nabla_\xi$,

$\mathbf{w}$ is the velocity of $S_{\alpha\beta}$,

$\mathbf{w}^c$ is the velocity of $S_{\alpha\beta}$ normal to $S_{\alpha\beta}$, $\mathbf{w}^c = \mathbf{nn} \cdot \mathbf{w}$,

$\mathbf{w}^s$ is the velocity of $S_{\alpha\beta}$ tangent to $S_{\alpha\beta}$, $\mathbf{w}^s = \mathbf{w} - \mathbf{w}^c$,
f is a surficial function $f(\mathbf{u}, t)$ defined on $S_{\alpha\beta}$,
$\mathbf{f}$ is a surficial vector $\mathbf{f}(\mathbf{u}, t)$ defined on $S_{\alpha\beta}$,
$\mathbf{f}^c$ is the vector normal to $S_{\alpha\beta}$ such that $\mathbf{f}^c = \mathbf{nn}\bullet\mathbf{f}$, and
$\mathbf{f}^s$ is the vector tangent to $S_{\alpha\beta}$ such that $\mathbf{f}^s = \mathbf{f} - \mathbf{f}^c$.

G[2, (0, 1), 2] Gradient

$$\int_{S_{\alpha\beta}} \nabla^s f dS = \int_{S_{\alpha\beta}} (\nabla^s \bullet \mathbf{n})\, \mathbf{n} f dS + \int_{C_{\alpha\beta}} \boldsymbol{\nu} f dS + \int_{C_{\alpha\beta\epsilon}} \boldsymbol{\nu} f dS \tag{8.65}$$

D[2, (0, 1), 2] Divergence

$$\int_{S_{\alpha\beta}} \nabla^s \bullet \mathbf{f} dS = \int_{S_{\alpha\beta}} (\nabla^s \bullet \mathbf{n})\, \mathbf{n} \bullet \mathbf{f}^c dS + \int_{C_{\alpha\beta}} \boldsymbol{\nu} \bullet \mathbf{f}^s dC + \int_{C_{\alpha\beta\epsilon}} \boldsymbol{\nu} \bullet \mathbf{f}^s dC \tag{8.66}$$

C[2, (0, 1), 2] Curl

$$\int_{S_{\alpha\beta}} \nabla^s \times \mathbf{f} dS = \int_{S_{\alpha\beta}} (\nabla^s \bullet \mathbf{n})\, \mathbf{n} \times \mathbf{f}^s dS + \int_{C_{\alpha\beta}} \boldsymbol{\nu} \times \mathbf{f} dC + \int_{C_{\alpha\beta\epsilon}} \boldsymbol{\nu} \times \mathbf{f} dC \tag{8.67}$$

T[2, (0, 1), 2] Transport

$$\int_{S_{\alpha\beta}} \left.\frac{\partial f}{\partial t}\right|_{\mathbf{u}} dS = \frac{\partial}{\partial t} \int_{S_{\alpha\beta}} f dS - \int_{S_{\alpha\beta}} (\nabla^s \bullet \mathbf{n})\, \mathbf{n} \bullet \mathbf{w}^c f dS - \int_{C_{\alpha\beta\epsilon}} \boldsymbol{\nu} \bullet \mathbf{w}^s f dC$$

$$+ \int_{C_{\alpha\beta_{\text{faces}}}} \frac{\mathbf{n}^* \bullet \mathbf{w} f}{\boldsymbol{\nu} \bullet \mathbf{n}^*} dC - \int_{C_{\alpha\beta}} \boldsymbol{\nu} \bullet \mathbf{w}^s f dC \tag{8.68}$$

8.6.3 Curvilineal Operator Theorems, [1, (0, 1), 2]

Figure and Notation

Refer to figure 8.9

$C_{\alpha\beta\epsilon}$ is the contact curve within the REV that is the location where the α- β- and ϵ-phases meet,

ext refers to the points of intersection of the $C_{\alpha\beta\epsilon}$ curve with the external boundary of the REV,

pts refers to the end points of the $C_{\alpha\beta\epsilon}$ within the REV where the

α- β- and ϵ-phases meet a fourth phase,

faces$_{\alpha\beta\epsilon}$ is the points on the faces of the REV where the bounding surface of the slab is intersected by $C_{\alpha\beta\epsilon}$,

n* is the unit vector normal to the external boundary of the REV pointing outward from the REV,

$\boldsymbol{\lambda}$ is the unit vector tangent to $C_{\alpha\beta\epsilon}$,

e is the unit vector tangent to $C_{\alpha\beta\epsilon}$ at pts and faces$_{\alpha\beta\epsilon}$ pointing outward from the curve such that $\mathbf{e}\cdot\boldsymbol{\lambda} = \pm 1$,

∇^c is the microscopic curvilineal operator, $\nabla^c = \boldsymbol{\lambda}\boldsymbol{\lambda}\cdot\nabla_\xi$,

w is the velocity of $C_{\alpha\beta\epsilon}$,

$\mathbf{w}^c$ is the velocity of $C_{\alpha\beta\epsilon}$ tangent to $C_{\alpha\beta\epsilon}$, $\mathbf{w}^c = \boldsymbol{\lambda}\boldsymbol{\lambda}\cdot\mathbf{w}$,

$\mathbf{w}^s$ is the velocity of $C_{\alpha\beta\epsilon}$ normal to $C_{\alpha\beta\epsilon}$, $\mathbf{w}^s = \mathbf{w} - \mathbf{w}^c$,

f is a curvilineal function $f(\mathbf{l}, t)$ defined on $C_{\alpha\beta\epsilon}$,

f is a curvilineal vector $\mathbf{f}(\mathbf{l}, t)$ defined on $C_{\alpha\beta\epsilon}$,

$\mathbf{f}^c$ is the vector tangent to $C_{\alpha\beta\epsilon}$ such that $\mathbf{f}^c = \boldsymbol{\lambda}\boldsymbol{\lambda}\cdot\mathbf{f}^c$, and

$\mathbf{f}^s$ is the vector normal to $C_{\alpha\beta\epsilon}$ such that $\mathbf{f}^s = \mathbf{f} - \mathbf{f}^c$.

G[1, (0, 1), 2] Gradient

$$\int_{C_{\alpha\beta\epsilon}} \nabla^c f dC = -\int_{C_{\alpha\beta\epsilon}} (\boldsymbol{\lambda}\cdot\nabla^c\boldsymbol{\lambda}) f dC + \sum_{\text{pts}} \mathbf{e}f + \sum_{\text{ext}} \mathbf{e}f \qquad \textbf{(8.69)}$$

D[1, (0, 1), 2] Divergence

$$\int_{C_{\alpha\beta\epsilon}} \nabla^c\cdot\mathbf{f} dC = -\int_{C_{\alpha\beta\epsilon}} (\boldsymbol{\lambda}\cdot\nabla^c\boldsymbol{\lambda})\cdot\mathbf{f}^s dC + \sum_{\text{pts}} \mathbf{e}\cdot\mathbf{f}^c + \sum_{\text{ext}} \mathbf{e}\cdot\mathbf{f}^c \qquad \textbf{(8.70)}$$

C[1, (0, 1), 2] Curl

$$\int_{C_{\alpha\beta\epsilon}} \nabla^c \times \mathbf{f} dC = -\int_{C_{\alpha\beta\epsilon}} (\boldsymbol{\lambda}\cdot\nabla^c\boldsymbol{\lambda}) \times \mathbf{f} dC + \sum_{\text{pts}} \mathbf{e}\times\mathbf{f}^s + \sum_{\text{ext}} \mathbf{e}\times\mathbf{f}^s \qquad \textbf{(8.71)}$$

T[1, (0, 1), 2] Transport

$$\int_{C_{\alpha\beta\epsilon}} \left.\frac{\partial f}{\partial t}\right|_1 dC = \frac{\partial}{\partial t}\int_{C_{\alpha\beta\epsilon}} f dC + \int_{C_{\alpha\beta\epsilon}} \boldsymbol{\lambda}\bullet\nabla^c\boldsymbol{\lambda}\bullet\mathbf{w}^s f dC - \sum_{\text{pts}} \mathbf{e}\bullet\mathbf{w}^c f$$

$$+ \sum_{\text{faces}_{\alpha\beta\epsilon}} \frac{\mathbf{n}^*\bullet\mathbf{w}f}{\mathbf{e}\bullet\mathbf{n}^*} - \sum_{\text{ext}} \mathbf{e}\bullet\mathbf{w}^c f \qquad \textbf{(8.72)}$$

8.7 CONCLUSION

The averaging theorems presented in this chapter have applicability to multiphase systems as well as in systems where it is desired to change the scale of modeling from the microscale to an intermediate macroscale and thereby filter out high frequency (in space) oscillations. Although none of the theorems involving the curl operator seem to have been presented explicitly in the literature, the form of the theorems is consistent operationally among gradient, divergence, and curl.

The theorems in the $[3, (3, 0), 0]$ family, equations (8.1) through (8.4), are widely known simply as the "averaging theorems" for porous media flow since they are readily seen to be applicable to facilitate a transformation from the microscale (or pore scale) to the macroscale (or continuum scale). *Slattery* [1967] and *Whitaker* [1967] developed the spatial averaging theorems by analogy with the transport theorem (i.e. the theorem in Chapter 7 designated as $T[3, (0, 0), 3]$). *Gray and Lee* [1976] presented a proof of the theorems that makes use of the spatial generalized function. In there works, these investigators have also obtained the theorems in the $[3, (0, 3), 0]$ family, equations (8.13) through (8.16), and shown that, in terms of the theorems presented here, the right sides of corresponding $[3, (3, 0), 0]$ and $[3, (0, 3), 0]$ theorems must be equal. This allows some supplementary information to be obtained.

The theorems from the $[2, (3, 0), 0]$ family, equations (8.5) through (8.8), have been presented previously by *Gray and Hassanizadeh* [1989]. In their work, the divergence form was presented for the special case of equation (8.6) where the surface vector $\mathbf{f}$ is tangent to the surface (i.e. $\mathbf{f}\bullet\mathbf{n} = 0$). These theorems are important for modeling the interfaces in a multiphase porous media system [*Hassanizadeh and Gray*, 1990; *Gray and Hassanizadeh*, 1991]. Also important for a complete study of a multiphase system would be an analysis of the contact lines. Such an analysis would make use of the theorems in the $[1, (3, 0), 0]$ family.

The only theorems in this chapter other than those in Sections 8.1 and 8.2 providing a transformation from the microscale to the macroscale that seem to appear elsewhere in the literature are those from the $[3, (2, 0), 1]$ family.

The divergence and transport theorems from this family were developed and used by *Gray* [1982] to obtain vertically averaged equations for multiphase flow in porous media. Indeed, the most useful of the averaging theorems would seem to be those from families that may be designated $[3,\ (m, 0), 3-m]$, for $m = 1, 2,$ or 3, since this set would be used in converting spatial balance equations at the microscale to balances that involve m macroscales and $3-m$ megascales. Nevertheless, the remaining theorem families, those for which the first index is not 3, will be useful in assessing the impact of surfaces and contact lines in multiphase systems at the macroscale.

8.8 REFERENCES

Gray, W. G., Derivation of vertically averaged equations describing multiphase flow in porous media, *Water Resources Research*, **18**(6), 1705-1712, 1982.

Gray, W. G., and P. C. Y. Lee, On the theorems for local volume averaging of multiphase systems, *Int. J. Multiphase Flow*, **3**, 333-340, 1977.

Gray, W. G., and S. M. Hassanizadeh, Unsaturated flow theory including interfacial phenomena, *Water Resources Research*, **27**(8), 1855-1863, 1991.

Hassanizadeh, S. M., and W. G. Gray, Mechanics and thermodynamics of multiphase flow in porous media including interface boundaries, *Advances in Water Resources*, **13**(4), 169-186, 1990.

Slattery, J. C., Flow of viscoelastic fluids through porous media, *AIChE Journal*, **13**, 1066-1071, 1967.

Whitaker, S., Diffusion and dispersion in porous media, *AIChE Journal*, **13**, 420-427, 1967.

Whitaker, S., Advances in theory of fluid motion in porous media, *Ind. Engng. Chem.*, **61**, 14-28, 1969.

CHAPTER NINE

APPLICATIONS

9.0 INTRODUCTION

The integration and averaging theorems that appear in Chapters 7 and 8 constitute a set of 128 theorems that have been derived using generalized functions. The current chapter will provide some sample applications of the theorems with examples chosen from engineering analysis.

Before proceeding with the examples, it is important to emphasize that the relations found in Chapters 7 and 8 are theorems applicable to physical problems for which a transformation between a point representation and a representation at some larger scale is sought. As with all theorems in mathematics, the tools used to prove each theorem and the logic behind the proof itself hold significant pedagogical value, but complete mastery of the proof is not a prerequisite for proper use of the theorem. Instead, the theorems presented in Chapters 7 and 8 provide appropriate tools to change the scale of a problem involving almost any geometry in one, two, or three dimensions. An apt analogy is that the collection of theorems in this work may be used much like a table of integrals so common in mathematics handbooks. The theorems are applicable to a wide variety of problems that are typically solved in only one or two dimensions and to problems that require some averaging because of complex microstructure.

Many engineering analyses are based on balance or conservation principles for mass, momentum, energy, and entropy. For some systems, it may be desirable, for example, to perform this balance at a microscopic point in space; for others, it may be desirable to consider a point on a surface; for still others, a large volume may be the appropriate scale at which to apply the balance equation. The integration and averaging theorems presented in this work provide a convenient set of rules that allow transformation from a point balance in three-dimensions to a point balance in fewer dimensions or to a balance at a larger scale. Because of their complexity, some systems are best modeled at a larger scale with microscopic phenomena accounted for by parametric methods. Thus by using integration and averaging theorems, one can transform point balance equations directly to an appropriate scale of study.

An alternative method for changing the scale of a problem, and one found frequently in the literature [e.g. *Abbot,* 1979; *Cunge et al.,* 1980; *Fetter,* 1988], is to develop balance laws at different scales starting from first principles. For example, the balance of mass in an estuary in which vertical variation in density is negligible and for which the pressure distribution in the vertical direction is hydrostatic can be obtained by analyzing flux and accumulation terms on a control volume that spans the depth of the fluid. This approach has the following shortcomings: the derivation of balance laws by examination of a geometrically complex element can be tedious; the procedure must be repeated for each system under study; important terms can be overlooked; and application to a system with complex geometry is difficult.

In contrast, use of integration and averaging theorems eliminates these problems because the point of departure is the microscale balance laws. Thus derivations at larger scales using complex geometric elements are avoided. Also, when the balance equations at the microscale are well-understood, all terms in the equations that represent important physical processes are included from the outset. Furthermore, curvature effects from arbitrary system geometries are incorporated naturally because they are accounted for in the theorems. Finally, use of the theorems generates macroscale/megascale terms that offer significant insight into the physics of the problem.

For any approach to a change of scale, appropriate constitutive theory at the larger scale is needed. This is referred to as a closure problem and is beyond the scope of this work. Therefore, the examples presented herein will be left in terms of the primitive variables that arise.

Salient features of the integration or averaging procedure will be illustrated in this chapter with six example applications, five chosen from hydrology and hydraulics, involving a variety of scales. Although most of the applications are drawn from the fluids literature, the methodology is applicable to problems from any discipline. The examples are: 1) conservation of mass for a fluid at the microscale; 2) conservation of mass for open channel flow; 3) conservation of momentum for open channel flow; 4) conservation of mass for an arbitrary surface; 5) conservation of mass for axisymmetric flow to a well in a porous medium with no vertical variation; and 6) derivation of additional theorems based on those tabulated.

The first example has been included to demonstrate the utility of the theorems in deriving microscale balance laws from first principles at the megascale. Examples 2 through 5 demonstrate the change in scale of microscopic balance laws using the appropriate theorems from Chapters 7 and 8. The resulting expressions are very general differential balance laws at some other scale. Some of the equations obtained will be shown to be generalized forms of expressions commonly found in the literature that simplify to those expressions under appropriate assumptions. Example 6 is chosen to emphasize that the theorems explicitly listed here are a formidable set of mathematical tools that, nevertheless, do not necessarily comprise the set of all useful relations for a change of scale. The approach of using the theorems given here to obtain additional forms is demonstrated as an alternative to obtaining additional forms by working with generalized functions.

9.1 CONSERVATION OF MASS FOR A FLUID AT THE MICROSCALE

In this example, the balance of mass for a fluid at the microscale will be derived as a mathematical expression of the observation that, under non-nuclear conditions, mass is neither created nor destroyed. Often, fluid mechanics texts present the control volume approach, an approach that first balances fluxes and accumulation on a cubic element and then examines the balance statement as the size of the element goes to zero [e.g. *Bird et al.*, 1960; *White*, 1979]. This method is often referred to as the Eulerian approach. As an alternative here, the theorems presented in Chapters 7 and 8 are used as tools to derive the balance of mass at the microscale, an approach often referred to as the Lagrangian approach [e.g. *Whitaker*, 1968; *John and Haberman*, 1980]. While the following derivation is certainly not unique to this work, it serves as a good introductory illustration of the use of integration and averaging theorems as powerful, yet simple, tools in engineering analysis.

Consider a fluid continuum that is allowed to deform as it translates through space. Define a mathematical volume, V, that is coincident with the fluid continuum such that the surface of the volume deforms with the fluid so that no fluid crosses the boundary surface of the volume. The mass of fluid in the volume, M, is given by:

$$M = \int_V \rho dV \tag{9.1}$$

where $\rho(\mathbf{x}, t)$ is the density of the fluid. Since no fluid crosses the boundary of the volume, the mass of fluid contained in V will be constant. Mathematically, this can be stated:

$$\frac{DM}{Dt} = \frac{D}{Dt}\int_V \rho dV = 0 \tag{9.2}$$

where D/Dt is the material time derivative, the derivative taken with respect to time following the volume. Equation (9.2) is a relation at the megascale since it describes mass conservation for the volume but says nothing about the distribution of mass within the volume. However, this equation can be converted to a microscale form by applying the appropriate integration theorems from the $[3, (0,0), 3]$ family. (Recall from Chapter 5 that the first index in this expressions indicates the number of microscopic dimensions of interest while the last index indicates the number of megascopic scales.)

Theorem $T[3, (0,0), 3]$, equation (7.4), is commonly referred to as the general transport theorem and provides a relation between the partial time derivative of a microscopic function holding spatial coordinates constant and the time derivative taken at the megascale. When density is the dependent variable, this theorem has the form:

$$\frac{d}{dt}\int_V \rho dV = \int_V \left.\frac{\partial\rho}{\partial t}\right|_{\mathbf{x}} dV + \int_S \rho \mathbf{w} \bullet \mathbf{n} dS \qquad \textbf{(9.3)}$$

where S is the boundary of the volume, $\mathbf{n}$ is the unit vector normal to S positive outward from V, $\mathbf{w}$ is the velocity of the boundary, and ρ is continuous within the volume. If no fluid is to leave or enter the volume under consideration, $\mathbf{w}\bullet\mathbf{n}$ must equal $\mathbf{v}\bullet\mathbf{n}$ on S where $\mathbf{v}$ is the velocity of the fluid. For this case, the total derivative in equation (9.3) is a material derivative and the equation takes the special form:

$$\frac{D}{Dt}\int_V \rho dV = \int_V \left.\frac{\partial\rho}{\partial t}\right|_{\mathbf{x}} dV + \int_S \rho \mathbf{v} \bullet \mathbf{n} dS \qquad \textbf{(9.4)}$$

which is the Reynolds transport theorem applied to density. Comparison of this equation to equation (9.2) indicates that both sides of the equation must be zero such that:

$$\int_V \left.\frac{\partial\rho}{\partial t}\right|_{\mathbf{x}} dV + \int_S \rho \mathbf{v} \bullet \mathbf{n} dS = 0 \qquad \textbf{(9.5)}$$

The divergence theorem, $D[3, (0,0), 3]$, equation (7.2), can be applied to the surface integral in equation (9.5) to convert it to a volume integral when $\rho\mathbf{v}$ is continuous within the volume to obtain:

$$\int_V \left.\frac{\partial\rho}{\partial t}\right|_{\mathbf{x}} dV + \int_V \nabla\bullet(\rho\mathbf{v}) dV = 0 \qquad \textbf{(9.6)}$$

Note that because equation (9.6) is valid for any arbitrary volume, the sum of the integrands at any point must be zero. Thus the microscopic equation for mass conservation without internal sources or sinks of mass is:

$$\frac{\partial \rho}{\partial t} + \nabla \cdot (\rho \mathbf{v}) = 0 \qquad \textbf{(9.7)}$$

where the explicit notation that the partial derivative is evaluated with spatial coordinates held constant has been dropped.

The first term in equation (9.7) accounts for the accumulation of mass and the second represents the net advective flux. Equation (9.7) will serve as the point of departure for many of the examples that follow.

9.2 CONSERVATION OF MASS FOR OPEN CHANNEL FLOW

The goal of this example is to derive the one-dimensional form of the mass balance equation that is applicable to open channel flow. As is the convention throughout this work, "one-dimensional" implies that the equation obtained will be valid for a general curve in space, not necessarily restricted to a straight line. The appropriate equation may be obtained by transforming the mass balance equation for a pure fluid from the fully microscopic form to a form that is megascopic in the directions normal to the flow direction and either macroscopic or microscopic along the channel axis. The first of these approaches uses the averaging theorems of Chapter 8 while the second uses the integration theorems of Chapter 7. As will be shown, both approaches lead to the same result.

9.2.1 Derivation Using Averaging Theorems

The first step in any application is the identification of the appropriate theorems to apply to the differential equation. In this subsection, an averaging approach is presented which means that the needed theorems may be found in Chapter 8. In fact, for the problem under consideration, the theorems to be used are members of the $[3, (1, 0), 2]$ in Section 8.5.1. This notation, explained in Chapter 5, indicates that the theorems transform a spatially microscopic equation to a form that accounts for macroscopic variation in one-dimension but is megascopic in the other two dimensions. The integration region to be used has been presented in figure 8.7, and the operator and coordinate conventions are those provided at the beginning of Section 8.5.1. The equation to be averaged, equation (9.7), contains both a partial time derivative and a divergence operator so that equations (8.52), $T[3, (1, 0), 2]$, and (8.50), $D[3, (1, 0), 2]$, respectively will be applied to these terms.

Application of transport theorem $T[3, (1, 0), 2]$, equation (8.52), to the first term in equation (9.7) yields:

$$\int_{V_\alpha} \frac{\partial \rho}{\partial t}\bigg|_{\mathbf{x}} dV = \frac{\partial}{\partial t}\int_{V_\alpha} \rho dV - \int_{S_{\alpha\beta}} \rho \mathbf{w} \bullet \mathbf{n} dS - \int_{S_{\alpha_{\text{edge}}}} \rho \mathbf{w} \bullet \mathbf{n}^* dS \tag{9.8}$$

This equation applies to a multiphase system where the α-phase is the phase of interest. For the case of open channel flow where only a single fluid phase is present, V_α is equal to V, the volume of the REV that intersects the river. Secondly, since no interface between phases is present, $S_{\alpha\beta}$ is zero. Finally, the last term in equation (9.8) will be an integral over the entire thin edge of the averaging region since only one phase is present so that $S_{\alpha_{\text{edge}}}$ may be written S_{edge}. Thus the required relation becomes:

$$\int_{V} \frac{\partial \rho}{\partial t}\bigg|_{\mathbf{x}} dV = \frac{\partial}{\partial t}\int_{V} \rho dV - \int_{S_{\text{edge}}} \rho \mathbf{w} \bullet \mathbf{n}^* dS \tag{9.9}$$

Application of divergence theorem $D[3, (1, 0), 2]$, equation (8.50), to the second term in equation (9.7) yields:

$$\int_{V} \nabla \bullet (\rho \mathbf{v})\, dV = \nabla^C \bullet \int_{V} \rho \mathbf{v}^C dV - \int_{V} \mathbf{N} \bullet \nabla^C \mathbf{N} \bullet \rho \mathbf{v} dV + \int_{S_{\text{edge}}} \rho \mathbf{v} \bullet \mathbf{n}^* dS \tag{9.10}$$

where the simplifications for a single phase system mentioned preceding equation (9.9) have been employed and $\mathbf{N}$ is the unit vector tangent to the channel axis.

The required averaged mass balance equation for an open channel is obtained by summation of equations (9.9) and (9.10) and invoking equation (9.6) such that:

$$\frac{\partial}{\partial t}\int_{V} \rho dV + \nabla^C \bullet \int_{V} \rho \mathbf{v}^C dV - \int_{V} \mathbf{N} \bullet \nabla^C \mathbf{N} \bullet \rho \mathbf{v} dV + \int_{S_{\text{edge}}} \rho (\mathbf{v} - \mathbf{w}) \bullet \mathbf{n}^* dS = 0 \tag{9.11}$$

On the left side of this equation, the first term accounts for the rate of change of mass within the REV; the second represents the net advective flux in the direction of the channel axis; the third term accounts for the effects of curvature; and the last term accounts for the flux of water across the edge of the REV due to mechanisms such as evaporation, precipitation, overland flow, and exchange of surface water with the groundwater.

The notation in equation (9.11) can be simplified after defining the following averages:

Intrinsic Volume Average

$$\langle G \rangle = \frac{1}{V}\int_V G dV \tag{9.12a}$$

Mass Average

$$\overline{G} = \frac{1}{\langle \rho \rangle V}\int_V \rho G dV \tag{9.12b}$$

Use of these definitions allows equation (9.11) to be written as:

$$\frac{\partial(\langle \rho \rangle V)}{\partial t} + \nabla^C \bullet (\langle \rho \rangle \bar{\mathbf{v}}^C V) - \mathbf{N} \bullet \nabla^C \mathbf{N} \bullet \langle \rho \rangle \bar{\mathbf{v}} V + \int_{S_{\text{edge}}} \rho(\mathbf{v} - \mathbf{w}) \bullet \mathbf{n}^* dS = 0 \tag{9.13}$$

For the slab REV used in this derivation, the volume is the product of the average cross sectional area, A, and the thickness of the slab, Δl, such that $V = A\Delta l$. Recall that for the slab averaging region, the thickness is required to be constant but the cross-sectional area can be a function of time and space. Furthermore the restriction that $|\mathbf{N} \bullet \nabla^C \mathbf{N}|\Delta l \ll 1$ has been assumed to hold. Accordingly, equation (9.13) can be divided by Δl to provide a general form of the mass balance equation for flow in an open channel:

$$\frac{\partial(\langle \rho \rangle A)}{\partial t} + \nabla^C \bullet (\langle \rho \rangle \bar{\mathbf{v}}^C A) - \mathbf{N} \bullet \nabla^C \mathbf{N} \bullet \langle \rho \rangle \bar{\mathbf{v}} A + \frac{1}{\Delta l}\int_{S_{\text{edge}}} \rho(\mathbf{v} - \mathbf{w}) \bullet \mathbf{n}^* dS = 0 \tag{9.14}$$

This equation can alternatively be cast into Lagrangian form by expanding the second term and rearranging to obtain:

$$\frac{D^C(\langle \rho \rangle A)}{Dt} + \langle \rho \rangle A \nabla^C \bullet \bar{\mathbf{v}}^C - \mathbf{N} \bullet \nabla^C \mathbf{N} \bullet \langle \rho \rangle \bar{\mathbf{v}} A + \frac{1}{\Delta l}\int_{S_{\text{edge}}} \rho(\mathbf{v} - \mathbf{w}) \bullet \mathbf{n}^* dS = 0 \tag{9.15}$$

where:

$$\frac{D^C}{Dt} = \frac{\partial}{\partial t} + \bar{\mathbf{v}}^C \bullet \nabla^C \tag{9.16}$$

In applying equation (9.14), one may find it convenient to break the boundary flux term given by the last integral into two terms:

$$\int_{S_{\text{edge}}} \rho(\mathbf{v}-\mathbf{w})\bullet\mathbf{n}^*\,dS = \int_{S_{\text{atm}}} \rho(\mathbf{v}-\mathbf{w})\bullet\mathbf{n}^*\,dS + \int_{S_{\text{bed}}} \rho(\mathbf{v}-\mathbf{w})\bullet\mathbf{n}^*\,dS \quad \textbf{(9.17)}$$

where:

S_{atm} is the portion of the REV edge exposed to the atmosphere, and

S_{bed} is the portion of the edge in contact with the earth.

Thus, the first integral on the right side of (9.17) accounts for the flux of water into or out of the channel at the upper surface due to processes such as evaporation, precipitation, and overland runoff. The second integral accounts for exchange of water in the channel with the adjacent aquifer (i.e. water bearing formation). Often, these flux integrals are indicated simply as source/sink type terms in the differential balance law. Here, and throughout this chapter, they will be left in integral form.

Frequently, fluid mechanics textbooks derive the mass and momentum conservation equation for open channel flow under a series of assumptions which are collectively referred to as the de St. Venant hypotheses [e.g., *Cunge et al.*, 1980]. They may be stated as:

1. The flow is one-dimensional, i.e., the velocity is uniform over the cross section and the water level across the section is horizontal.
2. The streamline curvature is small and vertical accelerations are negligible, hence the pressure distribution is hydrostatic.
3. The effects of boundary friction and turbulence can be accounted for through resistance laws analogous to those used for steady state flow.
4. The average channel bed slope is small so that the cosine of the angle between the bed and the horizontal is essentially 1.

Furthermore, often the exchange of water with the environment is neglected, and the density of the fluid is assumed to be constant. These assumptions will now be applied to equation (9.14) to simplify it to the form commonly employed in the literature.

When the exchange of water in the channel with the surrounding environment is negligible, the surface integral in equation (9.14) can be dropped. Also, the second of the de St. Venant hypotheses states that the streamline curvature is small. For this case, the third term in equation (9.14) involving $\mathbf{N}\bullet\nabla^C\mathbf{N}$ is negligible and may be neglected. When the average density, $\langle\rho\rangle$, is considered constant, the remaining terms in the equation may be divided by this parameter to produce:

$$\frac{\partial A}{\partial t} + \nabla^C \bullet (\bar{\mathbf{v}}^C A) = 0 \qquad \textbf{(9.18a)}$$

This equation may equivalently be written in the form most commonly found in the literature:

$$\frac{\partial A}{\partial t} + \frac{\partial (UA)}{\partial N} = 0 \qquad \textbf{(9.18b)}$$

where:

- U is the average velocity component along the channel axis, $\mathbf{N} \bullet \bar{\mathbf{v}}^C$, and
- N is the length coordinate along the channel.

Equation (9.18b) may be found in many texts dealing with open channel flow [e.g. *Cunge et al.*, 1980; *Jansen et al.*, 1979; *Le Méhauté*, 1976].

The advantage of the volumetric averaging approach presented above is that the method directly produces a very general form of the balance law using a series of simple steps. Then, if desired, simplifying assumptions can be invoked to reduce the general equation to a form used for a specific application. For example, channel curvature effects included in the general form of equation (9.14) are neglected by dropping the term containing $\mathbf{N} \bullet \nabla^C \mathbf{N}$. Retention of this term would require specification of the curvature along the channel axis obtained from field measurements as well as information about the velocity field in the directions orthogonal to the channel axis.

On the other hand, derivation of a balance law that applies in one or two dimensions based on a control volume approach can be difficult. The mathematical form that accounts for curvature is not readily apparent. It is easy to overlook some effect or be misled by intuition in formulating a balance expression for a curve. The averaging approach is more elegant, direct, and capable of revealing physical processes that come into play than direct construction of an equation for a moving control volume. The averaging approach also has the asset that it allows for an orderly progression from a complex expression to a simpler one under appropriate assumptions.

9.2.2 Derivation Using Integration Theorems

The mass balance equation for open channel flow can also be derived using the integration theorems from Chapter 7. The point of departure is the three-dimensional microscopic balance of mass given in equation (9.7). Also, since functional dependence in two megascopic directions will be removed by integration over a plane fixed in space, the appropriate choice of theorems will be the members of the $[3, (0,0), 2]$ family given in Section 7.2.2. This derivation is thus similar to that in Section 9.2.1 when the thickness of the averaging slab is zero.

Application of transport theorem $T[3, (0,0), 2]$, equation (7.20), to the time derivative in equation (9.7) and $D[3, (0,0), 2]$, equation (7.18), to the divergence term yields:

$$\frac{\partial}{\partial t}\int_A \rho dA + \nabla^c \bullet \int_A \rho \mathbf{v}^c dA - \int_A \mathbf{N} \bullet \nabla^c \mathbf{N} \bullet \rho \mathbf{v} dA + \int_C \frac{\rho(\mathbf{v}-\mathbf{w}) \bullet \mathbf{n}^*}{(\boldsymbol{\nu} \bullet \mathbf{n}^*)} dC = 0 \quad \textbf{(9.19)}$$

If definitions analogous to equations (9.12a) and (9.12b) are made for areal averages (i.e. volume averages in the limit where the thickness of the averaging volume is zero), equation (9.19) can be written as:

$$\frac{\partial \langle \rho \rangle A}{\partial t} + \nabla^c \bullet (\langle \rho \rangle \bar{\mathbf{v}}^c A) - \mathbf{N} \bullet \nabla^c \mathbf{N} \bullet \langle \rho \rangle \bar{\mathbf{v}} A + \int_C \frac{\rho(\mathbf{v}-\mathbf{w}) \bullet \mathbf{n}^*}{(\boldsymbol{\nu} \bullet \mathbf{n}^*)} dC = 0 \quad \textbf{(9.20)}$$

Note that in equation (9.20), the lower case "c" is indicative of the microscopic functional dependence along the channel axis, while in the equations from the last section, the capital "C" is indicative of macroscopic functional dependence along the axis since averaging over thickness Δl was utilized. Thus in the limit as the thickness $\Delta l \to 0$, equations (9.14) and (9.20) must be identical. For this to be the case, comparison of the equations shows that the following must hold:

$$\lim_{\Delta l \to 0} \left[\frac{1}{\Delta l} \int_{S_{\text{edge}}} \rho(\mathbf{v}-\mathbf{w}) \bullet \mathbf{n}^* dS \right] = \int_C \frac{\rho(\mathbf{v}-\mathbf{w}) \bullet \mathbf{n}^*}{(\boldsymbol{\nu} \bullet \mathbf{n}^*)} dC \quad \textbf{(9.21)}$$

To see that this expression is indeed valid, first recall that Δl is the thickness of the averaging slab. When the sides of the open channel are not orthogonal to the slab face such that the cross-section of the channel is not constant and $\mathbf{N} \bullet \mathbf{n}^* \neq 0$, the fact that the thickness of the boundary surface is larger than Δl must be taken into account. This increased thickness provides a larger element of area for interaction of the channel water with the surroundings than simply $\Delta l dC$. In fact, an element of the boundary surface, dS, will be obtained as $dC\Delta l/(\boldsymbol{\nu} \bullet \mathbf{n}^*)$. Thus equation (9.21) follows directly. The origin of the term $(\boldsymbol{\nu} \bullet \mathbf{n}^*)$ in the denominator of equation (9.20) was explained previously from a theoretical viewpoint in Section 4.6.2. The discussion in the current section provides a physical explanation.

One additional subtle difference between equations (9.14) and (9.20) does exist. In equation (9.14), averages that appear are over a volume, while in equation (9.20), the averages are over the cross-sectional area (Note that the size of the volume is an average cross-sectional area multiplied by Δl). The volumetric average of a quantity will be a somewhat smoother function of position along the channel axis than the areal average due to filtering of microscale fluctuations over Δl. This smoothing property of the averaging theorems is the reason they are used in developing equations for multiphase flow where it is desired to filter out small scale property changes due to phase interfaces. Corresponding average quantities in equations (9.14) and (9.20) will be equal only if Δl is of the same order of magnitude as the scale of microscale fluctuations.

Finally, note that equation (9.20) can be cast into the simplified mass balance form given as equation (9.18b) by invoking the de St. Venant hypotheses along with the assumptions of constant density and no sources/sinks of mass. Thus, the mass balance equation for single-phase open channel flow may be obtained using either integration or averaging theorems. Although these two procedures result in equations that are identical in form, the terms that appear must be interpreted differently unless $\Delta l \to 0$. In some applications where no microscale variation is to be modeled, the averaging theorems provide the appropriate theoretical tool for change of scale. General guidelines on the choice of theorems are expanded upon in the conclusion to this chapter.

9.3 CONSERVATION OF MOMENTUM FOR OPEN CHANNEL FLOW

In this example, the balance of momentum for open channel flow will be derived from the microscopic balance of momentum for a single phase fluid. Although the general microscale equation may be derived from the megascale equations in a manner analogous to the derivation of the mass balance equation in Section 9.1, this step will not be presented here. Momentum conservation is more complex than mass conservation because the mechanisms for transfer within the fluid must be taken into account. In particular, constitutive theory must be applied to relate the general stress tensor to pressure and the viscous stress. Derivations may be found in standard continuum mechanics or fluid mechanics books [e.g., *Malvern*, 1969; *Whitaker*, 1968]. The microscale momentum balance equation at a point in a fluid with continuous properties when gravity is the only body force is obtained as:

$$\frac{\partial \rho \mathbf{v}}{\partial t} + \nabla \cdot (\rho \mathbf{v}\mathbf{v}) - \rho \mathbf{g} - \nabla \cdot \mathbf{T} = 0 \qquad \textbf{(9.22)}$$

where:

- ρ is the fluid density,
- $\mathbf{v}$ is the fluid velocity,
- $\mathbf{g}$ is the gravity vector, and
- $\mathbf{T}$ is the total stress tensor.

If the fluid is considered to be a Stokesian fluid, the total stress tensor becomes:

$$\mathbf{T} = -p\mathbf{I} + \boldsymbol{\tau} \qquad \textbf{(9.23)}$$

where:

p is the pressure in the fluid,

$\mathbf{I}$ is the identity tensor, and

$\boldsymbol{\tau}$ is the viscous stress tensor.

Substitution of this expression into equation (9.22) yields the microscale momentum equation in the form:

$$\frac{\partial \rho \mathbf{v}}{\partial t} + \nabla \bullet (\rho \mathbf{v}\mathbf{v}) - \rho \mathbf{g} + \nabla p - \nabla \bullet \boldsymbol{\tau} = 0 \quad \textbf{(9.24)}$$

Equation (9.24) will be considered to be the turbulent form of the momentum balance such that the stress tensor accounts for viscous effects as well as Reynolds stresses.

As was illustrated with the mass balance equation, the balance of momentum for an open channel can be found by using either averaging or integration theorems. Here, only the derivation using averaging theorems will be presented. By reasoning similar to that presented in Section 9.2.1, the averaging theorems needed are in the $[3, (1, 0), 2]$ family of Section 8.5.1. Equation (9.24) contains partial time derivative, gradient and divergence operators and therefore the T, G, and D members of this theorem family will be needed. Note that $D[3, (1, 0), 2]$ is applicable to the divergence of a tensor (as well as a vector) as long as the order of the inner products is not commuted.

Integration of equation (9.24) over an averaging slab and application of the theorems provided in equations (8.49), (8.50), and (8.52) for a single phase system (i.e. no $S_{\alpha\beta}$ interfaces and $S_{\alpha_{\text{edge}}} = S_{\text{edge}}$ yields:

$$\frac{\partial}{\partial t}\int_V \rho \mathbf{v} dV + \nabla^C \bullet \int_V \rho \mathbf{v}^C \mathbf{v} dV - \int_V \mathbf{N} \bullet \nabla^C \mathbf{N} \bullet \rho \mathbf{v}\mathbf{v} dV - \int_V \rho \mathbf{g} dV + \nabla^C \int_V p dV$$

$$-\nabla^C \bullet \int_V \boldsymbol{\tau}^C dV + \int_V \mathbf{N} \bullet \nabla^C \mathbf{N} \bullet \boldsymbol{\tau} dV + \int_{S_{\text{edge}}} \mathbf{n}^* \bullet [\rho(\mathbf{v} - \mathbf{w})\mathbf{v} - \boldsymbol{\tau}] dS + \int_{S_{\text{edge}}} \mathbf{n}^* p dS = 0$$

(9.25)

where $\boldsymbol{\tau}^C \equiv \mathbf{N}\mathbf{N} \bullet \boldsymbol{\tau}$. For terms such as the first integral in the second line of equation (9.25) where the tensor is represented by a single entity, the definitions of average quantities given in equations (9.12a) and (9.12b) can be applied directly to obtain the macroscale relation. However, the averaging of the product of velocities, as in the second and third integrals in the first line of equation (9.25), is best accomplished when the microscopic velocity is expressed as a sum of a macroscopic mean velocity an a deviation term such that:

$$\mathbf{v}\,(N+n,\mathbf{u}_\xi,t) \;=\; \bar{\mathbf{v}}\,(N,t) + \tilde{\mathbf{v}}\,(N,\boldsymbol{\xi},t) \qquad \textbf{(9.26a)}$$

where:

- $\mathbf{v}\,(N+n,\mathbf{u}_\xi,t)$ is the microscale velocity,
- $\bar{\mathbf{v}}\,(N,t)$ is the mass average velocity for the REV,
- $\tilde{\mathbf{v}}\,(N,\boldsymbol{\xi},t)$ is the microscale deviation from $\bar{\mathbf{v}}\,(N,t)$ within the REV,
- N is the macroscale coordinate,
- n is the microscale coordinate in direction N within the REV,
- $\mathbf{u}_\xi$ are the two microscale coordinates orthogonal to n, and
- $\boldsymbol{\xi}$ indicates the three microscale coordinates.

The concepts behind selection of the dependence of the microscale function on the macroscale and microscale coordinates has been discussed previously in Sections 4.2 and 4.3. Here, because megascopic averaging is done over the directions normal to the channel axis, no dependence on macroscopic coordinates in those directions appears. Mass averaging of equation (9.26a) shows that the average of the term on the left is equal to the average of the first term on the right so that:

$$\bar{\tilde{\mathbf{v}}} = 0 \qquad \textbf{(9.26b)}$$

Note that since $\mathbf{v}^C = \mathbf{N}\mathbf{N}\bullet\mathbf{v}$ has the same functional dependence on macroscopic and microscopic coordinates as $\mathbf{v}$, equations similar to equations (9.26a) and (9.26b) apply to this function as well.

Since $\mathbf{N}\bullet\nabla^C\mathbf{N}$ depends only on the macroscopic coordinate and integration is over microscopic coordinates, this quantity may be moved outside the integral of the third term in equation (9.25). Use of the decomposition given by equation (9.26a) in this term provides:

$$\int_V \rho\mathbf{v}\mathbf{v}\,dV = \int_V \rho\,(\bar{\mathbf{v}}+\tilde{\mathbf{v}})\,(\bar{\mathbf{v}}+\tilde{\mathbf{v}})\,dV = \int_V \rho\,(\bar{\mathbf{v}}\bar{\mathbf{v}}+\tilde{\mathbf{v}}\tilde{\mathbf{v}})\,dV \qquad \textbf{(9.27)}$$

where the terms involving $\bar{\mathbf{v}}\tilde{\mathbf{v}}$ and $\tilde{\mathbf{v}}\bar{\mathbf{v}}$ are zero by equation (9.26b). A similar decomposition applies to the second integral in equation (9.25) so that this equation may be written as:

$$\frac{\partial(\langle\rho\rangle\bar{\mathbf{v}}V)}{\partial t} + \nabla^C\bullet(\langle\rho\rangle\bar{\mathbf{v}}^C\bar{\mathbf{v}}V) - \mathbf{N}\bullet\nabla^C\mathbf{N}\bullet\langle\rho\rangle\bar{\mathbf{v}}\bar{\mathbf{v}}V - \langle\rho\rangle\bar{\mathbf{g}}V + \nabla^C(\langle p\rangle V)$$
$$-\nabla^C\bullet[(\langle\boldsymbol{\tau}^C\rangle - \langle\rho\rangle\overline{\tilde{\mathbf{v}}^C\tilde{\mathbf{v}}})V] + \mathbf{N}\bullet\nabla^C\mathbf{N}\bullet[(\langle\boldsymbol{\tau}\rangle - \langle\rho\rangle\overline{\tilde{\mathbf{v}}\tilde{\mathbf{v}}})V]$$
$$+\int_{S_{\text{edge}}}\mathbf{n}^*\bullet(\rho(\mathbf{v}-\mathbf{w})\mathbf{v}-\boldsymbol{\tau})\,dS + \int_{S_{\text{edge}}}\mathbf{n}^*p\,dS = 0 \qquad \textbf{(9.28)}$$

where the averaging symbols of equations (9.12a) and (9.12b) have been employed.

The terms grouped together in the second line of equation (9.28) account for the apparent viscous stress at the macroscale. Considerable insight into the behavior of the fluid at the macroscale is found by examining these terms. At the macroscale, momentum interactions between adjacent fluid particles are due to average microscale surface stresses, represented by $\langle \boldsymbol{\tau}^C \rangle$ and $\langle \boldsymbol{\tau} \rangle$, and by the average of the microscale variations of the fluid velocity within the averaging volume arising from the inertial terms as $\langle \rho \rangle \overline{\tilde{\mathbf{v}}^C \tilde{\mathbf{v}}}$ and $\langle \rho \rangle \overline{\tilde{\mathbf{v}} \tilde{\mathbf{v}}}$. As alluded to in the introduction, the form of the stress terms, including the effects of curvature, arises naturally when macroscale balance laws are derived using volumetric averaging procedures. However, when balance equations are proposed at the macroscale from conservation principles, this type of information may be overlooked or implicitly buried in constitutive assumptions invoked during the derivation.

For the slab averaging region used in this example, the volume, V, equals the constant slab thickness, Δl, multiplied by the average area of the parallel faces, A. Substitution of $A \Delta l$ for V in equation (9.28) and division of the equation by Δl yields the macroscopic momentum balance:

$$\frac{\partial(\langle \rho \rangle \bar{\mathbf{v}} A)}{\partial t} + \nabla^C \cdot (\langle \rho \rangle \bar{\mathbf{v}}^C \bar{\mathbf{v}} A) - \mathbf{N} \cdot \nabla^C \mathbf{N} \cdot \langle \rho \rangle \bar{\mathbf{v}} \bar{\mathbf{v}} A - \langle \rho \rangle \bar{\mathbf{g}} A + \nabla^C (\langle p \rangle A)$$

$$- \nabla^C \cdot [(\langle \boldsymbol{\tau}^C \rangle - \langle \rho \rangle \overline{\tilde{\mathbf{v}}^C \tilde{\mathbf{v}}}) A] + \mathbf{N} \cdot \nabla^C \mathbf{N} \cdot [(\langle \boldsymbol{\tau} \rangle - \langle \rho \rangle \overline{\tilde{\mathbf{v}} \tilde{\mathbf{v}}}) A]$$

$$+ \frac{1}{\Delta l} \int_{S_{\text{edge}}} \mathbf{n}^* \cdot [\rho (\mathbf{v} - \mathbf{w}) \mathbf{v} - \boldsymbol{\tau}] \, dS + \frac{1}{\Delta l} \int_{S_{\text{edge}}} \mathbf{n}^* p \, dS = 0 \qquad \textbf{(9.29)}$$

where:

- $\mathbf{N}$ is the macroscopic unit vector tangent to the axis of the channel,
- ∇^C is the macroscopic spatial operator along the channel axis, $\mathbf{NN} \cdot \nabla$,
- $\langle \rho \rangle$ is the average fluid density within the REV,
- $\bar{\mathbf{v}}$ is the mass average fluid velocity within the REV,
- $\bar{\mathbf{v}}^C$ is the mass average fluid velocity along the channel axis, $\mathbf{NN} \cdot \bar{\mathbf{v}}$,
- A is the average cross-sectional area of the channel for the REV,
- $\bar{\mathbf{g}}$ is the mass averaged gravity vector,
- $\langle \boldsymbol{\tau} \rangle$ is the average turbulent stress tensor,
- $\langle \boldsymbol{\tau}^C \rangle$ is equal to $\mathbf{NN} \cdot \langle \boldsymbol{\tau} \rangle$, and
- $\langle p \rangle$ is the average pressure within the REV.

The first term in equation (9.29) accounts for the accumulation of momentum, the second is the momentum flux, the third and seventh account for effects due to channel curvature, the fourth is the body force term, the fifth and ninth terms account for net force due to fluid pressure, the sixth accounts for the effects of stress within the fluid, and the eighth accounts for momentum changes due to momentum influx at the edges of the averaging volume due to fluid inflow and frictional effects at the water surface and along the channel boundary. To provide a closed description of the momentum balance, the stresses (both internal and on the boundary) must be related to other dependent variables by invoking an appropriate constitutive theory. However, as this departs from the objective of the example, it will not be addressed here.

It is important to note that averaging of the momentum equation over the cross section of the channel does not change the character of the equation from a vector form with components in three orthogonal directions. The momentum balance in the direction of the channel axis, the primary direction of flow, can be obtained by forming the dot product of equation (9.29) with $\mathbf{N}$.

The de St. Venant hypotheses of Section 9.2.1 along with the assumption of constant density will now be used to further simplify equation (9.29) to a form often presented in the literature. The assumption of constant density implies that $\langle\rho\rangle = \rho$. Furthermore, equation (9.29) can be divided by the constant density to eliminate this quantity from most of the terms. Since the pressure distribution is hydrostatic by the second of the de St. Venant hypotheses, significant simplification of the pressure terms in equation (9.29) may be achieved. Let ζ be a coordinate in the vertical direction that is positive upward and $\zeta = H$ identify the location of the free surface of the channel. The gravity may be defined in terms of a potential function, $\phi = g\zeta$, such that:

$$\mathbf{g} = -\nabla\phi \tag{9.30a}$$

The pressure field will then be given by:

$$p = \rho g H - \rho\phi \tag{9.30b}$$

When the density is constant, the volume and mass averages given respectively by equations (9.12a) and (9.12b) are equal. Therefore substitution of equations (9.30a) and (9.30b) into the body force term and the terms containing pressure in equation (9.29) yields:

$$-\langle\rho\rangle\bar{\mathbf{g}}A + \nabla^C(\langle p\rangle A) + \frac{1}{\Delta l}\int\limits_{S_{\text{edge}}} \mathbf{n}^* p\,dS = \frac{1}{\Delta l}\int\limits_V \rho\nabla\phi\,dV + \nabla^C(\langle\rho g H\rangle A)$$
$$-\nabla^C\left(\frac{1}{\Delta l}\int\limits_V \rho\phi\,dV\right) + \frac{1}{\Delta l}\int\limits_{S_{\text{edge}}} \mathbf{n}^*\rho g H\,dS - \frac{1}{\Delta l}\int\limits_{S_{\text{edge}}} \mathbf{n}^*\rho\phi\,dS \tag{9.31}$$

By equation (8.49), theorem $G[3,\ (1,0),2]$, the terms containing ϕ in equation (9.31) cancel. If contributions of variation in H across the thickness of the averaging slab contribute negligibly to the equality in equation (9.31), this equation reduces to:

$$-\langle\rho\rangle\bar{\mathbf{g}}A + \nabla^C(\langle p\rangle A) + \frac{1}{\Delta l}\int_{S_{\text{edge}}} \mathbf{n}^* p\,dS = \nabla^C(\rho g H A) + \frac{\rho g H}{\Delta l}\int_{S_{\text{edge}}} \mathbf{n}^*\,dS \tag{9.32}$$

Now by equation (8.49) with $f = 1$, one obtains the identity:

$$\frac{1}{\Delta l}\int_{S_{\text{edge}}} \mathbf{n}^*\,dS = -\nabla^C A \tag{9.33}$$

so that equation (9.32) may be written:

$$-\langle\rho\rangle\bar{\mathbf{g}}A + \nabla^C(\langle p\rangle A) + \frac{1}{\Delta l}\int_{S_{\text{edge}}} \mathbf{n}^* p\,dS = \rho g A \nabla^C H \tag{9.34}$$

Substitution of this expression back into equation (9.29) and division by the constant density yields:

$$\frac{\partial(\bar{\mathbf{v}}A)}{\partial t} + \nabla^C\bullet(\bar{\mathbf{v}}^C\bar{\mathbf{v}}A) - \mathbf{N}\bullet\nabla^C\mathbf{N}\bullet\bar{\mathbf{v}}\bar{\mathbf{v}}A + gA\nabla^C H$$

$$-\frac{1}{\rho}\nabla^C\bullet[(\langle\boldsymbol{\tau}^C\rangle - \rho\overline{\tilde{\mathbf{v}}^C\tilde{\mathbf{v}}})A] + \frac{1}{\rho}\mathbf{N}\bullet\nabla^C\mathbf{N}\bullet[(\langle\boldsymbol{\tau}\rangle - \rho\overline{\tilde{\mathbf{v}}\tilde{\mathbf{v}}})A]$$

$$+\frac{1}{\rho\Delta l}\int_{S_{\text{edge}}} \mathbf{n}^*\bullet[\rho(\mathbf{v}-\mathbf{w})\mathbf{v} - \boldsymbol{\tau}]\,dS = 0 \tag{9.35}$$

The component of this equation along the channel axis may be obtained by taking the dot product with the unit vector $\mathbf{N}$. This provides the equation:

$$\frac{\partial(UA)}{\partial t} + \frac{\partial(UUA)}{\partial N} - 2U\mathbf{N}\bullet\nabla^C\mathbf{N}\bullet\bar{\mathbf{v}}A + gA\frac{\partial H}{\partial N}$$

$$-\frac{1}{\rho}\frac{\partial[(\mathbf{N}\bullet\langle\boldsymbol{\tau}^C\rangle\bullet\mathbf{N} - \rho\mathbf{N}\bullet\overline{\tilde{\mathbf{v}}\tilde{\mathbf{v}}}\bullet\mathbf{N})A]}{\partial N} + \frac{1}{\rho}\nabla^C\mathbf{N}:[(\langle\boldsymbol{\tau}\rangle - \rho\overline{\tilde{\mathbf{v}}\tilde{\mathbf{v}}})A]$$

$$+\frac{1}{\rho}\mathbf{N}\bullet\nabla^C\mathbf{N}\bullet[(\langle\boldsymbol{\tau}\rangle-\rho\overline{\bar{\mathbf{v}}\bar{\mathbf{v}}})A]\bullet\mathbf{N}+\frac{1}{\rho\Delta l}\int_{S_{\text{edge}}}\mathbf{n}^*\bullet[\rho(\mathbf{v}-\mathbf{w})\mathbf{v}-\boldsymbol{\tau}]\bullet\mathbf{N}dS = 0$$

(9.36)

where U is the velocity along the channel, as in equation (9.18b), and use is made of the fact that:

$$\nabla^C\bullet\mathbf{f}^C \equiv \frac{\partial(\mathbf{f}^C\bullet\mathbf{N})}{\partial N} \tag{9.37}$$

Now if the de St. Venant assumption is applied that channel curvature effects are small, all terms in equation (9.36) involving the gradient of the normal, $\nabla^C\mathbf{N}$, may be dropped to obtain:

$$\frac{\partial(UA)}{\partial t}+\frac{\partial(UUA)}{\partial N}+gA\frac{\partial H}{\partial N}-\frac{1}{\rho}\frac{\partial[(\mathbf{N}\bullet\langle\boldsymbol{\tau}^C\rangle\bullet\mathbf{N}-\rho\mathbf{N}\bullet\overline{\bar{\mathbf{v}}\bar{\mathbf{v}}}\bullet\mathbf{N})A]}{\partial N}$$

$$+\frac{1}{\rho\Delta l}\int_{S_{\text{edge}}}\mathbf{n}^*\bullet[\rho(\mathbf{v}-\mathbf{w})\mathbf{v}-\boldsymbol{\tau}]\bullet\mathbf{N}dS = 0 \tag{9.38}$$

The last two terms on the right side of this equation account, respectively, for stress effects within the fluid and momentum transfer at the boundary. Commonly, the internal stresses and stress at the interface between the water and the atmosphere are considered negligible in comparison to the stress exerted by the channel bed. If the momentum transfer to the channel in direction **N** due to infiltration and runoff is negligible such that over the boundary $\mathbf{n}^*\bullet(\mathbf{v}-\mathbf{w})\,\mathbf{v}\bullet\mathbf{N}$ is negligible, equation (9.38) may be written:

$$\frac{\partial(UA)}{\partial t}+\frac{\partial(UUA)}{\partial N}+gA\frac{\partial H}{\partial N}+\frac{P^w}{\rho}\tau_b = 0 \tag{9.39}$$

where:

P^w is the average wetted perimeter of the channel cross section, and

τ_b is the average stress exerted by the channel on the flow due to frictional resistance such that:

$$\tau_b = -\frac{1}{P^w\Delta l}\int_{P^w\Delta l}\mathbf{n}^*\bullet\boldsymbol{\tau}\bullet\mathbf{N}dS \tag{9.40}$$

where $P^w\Delta l$ is the wetted area of the stream bed in the averaging slab. Equation (9.39) is the momentum equation for an open channel under the restrictions of the de St. Venant hypotheses supplemented with the assumptions of constant density and negligible sources of momentum due to flows across the boundary of the channel cross-section. This equation, or some slight variant, is the form of the momentum balance for open channel flow frequently listed in the literature [e.g. *Cunge et al.*, 1980; *Jansen et al.*, 1979].

This discussion highlights the assumptions that must be introduced to obtain equation (9.39) from the general equation (9.29). Insight to the physics and the actual assumptions employed can be lost by attempting to derive equation (9.39) directly using a control volume that is megascopic for the channel cross section and macroscopic in the direction normal to the channel axis. With the averaging approach, all effects accounted for at the microscale are directly transformed to the macroscale in the general equation (9.29). Simplifications are made by eliminating terms from the general equation. With the control volume approach, one is faced with the prospect of accounting for complexities by somewhat arbitrarily adding terms to a relatively simple equation. Thus the averaging procedure admits the theoretically pleasing approach of moving from the complex to the simple by identification of assumptions rather than moving from the simple to the complex through somewhat heuristic addition of terms and parameters. For example, the correct form of the terms accounting for curvature arise naturally in equation (9.29) from the averaging approach. Direct postulation of the forms these terms take based on a control volume approach is difficult, at best.

9.4 CONSERVATION EQUATIONS FOR A SURFACE

In areas as diverse as multiphase flow and structural analysis of plates and shells, access to equations that describe the dynamics of a deforming and arbitrarily shaped surface is useful. The development of surface flow equations, for example, as in *Scriven* [1960] or *Aris* [1962] is complex and relies on extensive knowledge of differential geometry and metric tensors. Within the framework of the integration theorems, in particular the $[3,\ (0,0)\,,1]$ family, the surface equations may be developed in a way that is both physically satisfying and mathematically rigorous. In this section, both the mass and momentum balance equations will be developed.

9.4.1 Conservation of Mass

Here, the mass balance equation for a general surface within a fluid will be developed from the three-dimensional microscopic equation. Application to an interface between the bulk phases of a multiphase system will then be demonstrated.

The point of departure for this derivation is the balance of mass for a single phase fluid as given by equation (9.7). The family of theorems for integration of this equation over a straight line segment, equations (7.37) through (7.40) of Section 7.3.2, is applied to a thin layer of the fluid. The result of this process is a balance equation for the fluid layer that, in the limit of integration over a short line segment, is the balance equation for a surface.

Integration of equation (9.7) over a straight line segment of length ΔL fixed in a body of fluid gives:

$$\int_{\Delta L} \frac{\partial \rho}{\partial t} dL + \int_{\Delta L} \nabla \bullet (\rho \mathbf{v})\, dL = 0 \qquad \textbf{(9.41)}$$

Since the integration is of a three-dimensional point equation over one dimension, the $[3\,(0,0)\,,1]$ family of theorems is appropriate. In particular, integration of a time derivative and a divergence operator over a straight line fixed in space makes use of theorems $T[3,\ (0,0)\,,1]$ and $D[3,\ (0,0)\,,1]$, equations (7.40) and (7.38), respectively. Application of these theorems to equation (9.41) yields:

$$\frac{\partial}{\partial t}\int_{\Delta L} \rho dL + \nabla^s \bullet \int_{\Delta L} \rho \mathbf{v}^s dL + \int_{\Delta L} (\nabla \bullet \mathbf{\Lambda})\, \mathbf{\Lambda} \bullet \rho \mathbf{v} dL + \left[\frac{\rho(\mathbf{v}-\mathbf{w})\bullet\mathbf{n}^*}{(\mathbf{e}\bullet\mathbf{n}^*)}\right]\Bigg|_{\text{ends}} = 0$$

(9.42)

In the last term of this equation, it is very important to understand that **w** is the velocity of the ends of the line segment and **v** is the velocity of fluid entering or leaving at the end of the segment. The integrals in equation (9.42) may be expressed in terms of average values of the integrands multiplied by ΔL. If notation for averages similar to that used in equations (9.12a) and (9.12b) is introduced, with the integration region being ΔL rather than V, equation (9.42) becomes:

$$\frac{\partial}{\partial t}(\langle\rho\rangle\Delta L) + \nabla^s \bullet (\langle\rho\rangle\bar{\mathbf{v}}^s \Delta L) + (\nabla \bullet \mathbf{\Lambda})\, \mathbf{\Lambda} \bullet \rho \bar{\mathbf{v}} \Delta L + \left[\frac{\rho(\mathbf{v}-\mathbf{w})\bullet\mathbf{n}^*}{(\mathbf{e}\bullet\mathbf{n}^*)}\right]\Bigg|_{\text{ends}} = 0$$

(9.43)

Now define the mass per unit surface area as:

$$\Gamma = \langle\rho\rangle\Delta L \qquad \textbf{(9.44)}$$

This quantity is a measure of the mass contained in the surface of thickness ΔL. Substitution of this expression into equation (9.43) yields:

$$\frac{\partial \Gamma}{\partial t} + \nabla^s \bullet (\Gamma \bar{\mathbf{v}}^s) + (\nabla \bullet \mathbf{\Lambda})\, \mathbf{\Lambda} \bullet \Gamma \bar{\mathbf{v}} + \left[\frac{\rho(\mathbf{v}-\mathbf{w})\bullet\mathbf{n}^*}{(\mathbf{e}\bullet\mathbf{n}^*)}\right]\Bigg|_{ends} = 0 \qquad \textbf{(9.45)}$$

For the situation that ΔL approaches zero, the following limits are obtained:

$$\lim_{\Delta L \to 0} \mathbf{e} = \mathbf{n}^* \qquad \textbf{(9.46a)}$$

$$\lim_{\Delta L \to 0} \bar{\mathbf{v}} = \mathbf{w} \qquad \textbf{(9.46b)}$$

$$\lim_{\Delta L \to 0} \mathbf{n}^* = \pm\boldsymbol{\Lambda} \qquad \textbf{(9.46c)}$$

$$\lim_{\Delta L \to 0} [\mathbf{f} \bullet \mathbf{n}^*]|_{\text{ends}} = [\![\mathbf{f}]\!] \bullet \boldsymbol{\Lambda} \qquad \textbf{(9.46d)}$$

where $[\![\mathbf{f}]\!]$ indicates the jump in **f** when crossing the surface. Therefore, in the limit of integration over a surface of zero thickness, equation (9.45) may be written:

$$\frac{\partial \Gamma}{\partial t} + \nabla^s \bullet (\Gamma \mathbf{w}^s) + (\nabla^s \bullet \boldsymbol{\Lambda})\, \boldsymbol{\Lambda} \bullet \Gamma \mathbf{w} + [\![\rho(\mathbf{v} - \mathbf{w})]\!] \bullet \boldsymbol{\Lambda} = 0 \qquad \textbf{(9.47)}$$

where:

- $\mathbf{w}$ is the velocity of the fluid in the surface,
- $\mathbf{w}^s$ is velocity of the surface fluid in the direction tangent to the surface, $\mathbf{v}^s = \mathbf{v} - \boldsymbol{\Lambda}\boldsymbol{\Lambda} \bullet \mathbf{v}$, and
- $\mathbf{v}$ is the velocity of the bulk fluid adjacent to the surface.

In equation (9.47), the fact that $\nabla \bullet \boldsymbol{\Lambda} = \nabla^s \bullet \boldsymbol{\Lambda}$ has been employed.

Equation (9.47) governs the balance of mass for an arbitrary surface within a body of fluid. The first term accounts for the accumulation of mass in the surface, the second describes the net advective flux of mass within the surface, the third term accounts for mass changes due to curvature and movement of the surface in the direction normal to the surface, and the last term accounts for mass exchange with the fluid on both sides of the surface.

One application of equation (9.47) is to an interface between two phases of a multiphase system. In many systems, the thickness of the interface is small compared to the dimensions of the phases themselves so that the interface may be regarded as a surface. Frequently, no properties such as mass, momentum, or internal energy are assigned to this surface although these properties are exchanged across the surface by the surrounding fluids. However, the result presented in equation (9.47) provides a more general governing mass balance expression for an interface that may contain mass. In the special case where the interface is assumed to be massless, this equation reduces to:

$$[\![\rho(\mathbf{v} - \mathbf{w})]\!] \bullet \boldsymbol{\Lambda} = 0 \qquad \textbf{(9.48)}$$

which is the frequently used jump condition employed in descriptions of multiphase systems or of mass balance across singular surfaces [e.g. in *Eringen*, 1980; *Slattery*, 1972; *Bowen*, 1990]. Equation (9.48) states that for a massless interface, whatever mass leaves one fluid phase immediately enters the adjacent fluid phase.

9.4.2 Conservation of Momentum

The momentum balance equation for a surface may be developed by applying the $[3\,(0,0),1]$ family of theorems to the microscopic spatial momentum balance equation (9.22). The reasoning behind this derivation, particularly the idea of integrating over a straight line segment that is allowed to approach a length of zero in the limit, is the same as was used in Subsection 9.4.1 for the conservation of mass equation. Therefore, the present derivation will be presented in an abridged format.

As with the mass balance equation, the time derivative and divergence operators appear in this equation such that the theorems $T[3,\ (0,0),1]$ and $D[3,\ (0,0),1]$, equations (7.40) and (7.38), respectively, are needed. Application of these theorems to the point momentum balance for integration over a straight line segment of length ΔL yields:

$$\frac{\partial}{\partial t}\int\limits_{\Delta L}\rho\mathbf{v}dL+\nabla^s\cdot\int\limits_{\Delta L}\rho\mathbf{v}^s\mathbf{v}dL+\int\limits_{\Delta L}(\nabla\cdot\boldsymbol{\Lambda})\,\boldsymbol{\Lambda}\cdot\rho\mathbf{v}\mathbf{v}dL-\int\limits_{\Delta L}\rho\mathbf{g}dL$$

$$-\nabla^s\cdot\int\limits_{\Delta L}\mathbf{T}^{sx}dL-\int\limits_{\Delta L}(\nabla\cdot\boldsymbol{\Lambda})\,\boldsymbol{\Lambda}\cdot\mathbf{T}dL+\left[\frac{\rho\mathbf{v}\,(\mathbf{v}-\mathbf{w})\cdot\mathbf{n}^*-\mathbf{n}^*\cdot\mathbf{T}}{\mathbf{e}\cdot\mathbf{n}^*}\right]\Bigg|_{\text{ends}}=0 \tag{9.49}$$

where $\mathbf{T}^{sx}=\mathbf{T}-\boldsymbol{\Lambda}\boldsymbol{\Lambda}\cdot\mathbf{T}$ and the other symbols have been used previously in the derivation of the last subsection. Employment of the notation for averages introduced in equations (9.12a) and (9.12b) with the averaging region being ΔL, and use of the definition of mass per unit surface area as in equation (9.44) as $\Gamma=\rho\Delta L$ reduces equation (9.49) to the form:

$$\frac{\partial(\Gamma\bar{\mathbf{v}})}{\partial t}+\nabla^s\cdot(\Gamma\bar{\mathbf{v}}^s\bar{\mathbf{v}})+(\nabla\cdot\boldsymbol{\Lambda})\,\boldsymbol{\Lambda}\cdot\Gamma\bar{\mathbf{v}}\bar{\mathbf{v}}-\Gamma\mathbf{g}$$

$$-\nabla^s\cdot\mathbf{S}^{sx}-(\nabla\cdot\boldsymbol{\Lambda})\,\boldsymbol{\Lambda}\cdot\mathbf{S}+\left[\frac{\rho\mathbf{v}\,(\mathbf{v}-\mathbf{w})\cdot\mathbf{n}^*-\mathbf{n}^*\cdot\mathbf{T}}{\mathbf{e}\cdot\mathbf{n}^*}\right]\Bigg|_{\text{ends}}=0 \tag{9.50}$$

where:

S is the stress per unit surface area integrated through the thickness of the surface, defined as $\mathbf{S}=(\langle\mathbf{T}\rangle-\langle\rho\rangle\bar{\tilde{\mathbf{v}}}\bar{\tilde{\mathbf{v}}})\,\Delta L$.

In the limit as $\Delta L \to 0$, the last term on the left side accounts for the effects of the fluid phases on each side of the interface. This can be expressed as a jump condition, as was done for the mass balance in obtaining equation (9.47) from equation (9.45), so that the momentum balance equation for an interface is:

$$\frac{\partial (\Gamma \mathbf{w})}{\partial t} + \nabla^s \bullet (\Gamma \mathbf{w}^s \mathbf{w}) + (\nabla \bullet \Lambda)\, \Lambda \bullet \Gamma \mathbf{w}\mathbf{w} - \Gamma \mathbf{g}$$

$$- \nabla^s \bullet \mathbf{S}^{sx} - (\nabla \bullet \Lambda)\, \Lambda \bullet \mathbf{S} + [\![\rho \mathbf{v}(\mathbf{v} - \mathbf{w}) - \mathbf{T}]\!] \bullet \Lambda = 0 \tag{9.51}$$

where equations (9.46a) through (9.46d) have been invoked and the stress tensor for the fluid phases, $\mathbf{T}$, has been taken to be symmetric.

Although the general form of the momentum balance for a surface given by equation (9.51) is interesting in its own right, two special cases are of particular interest. First, when the interface is considered to be massless and also incapable of sustaining any stress, the momentum balance reduces to:

$$[\![\rho \mathbf{v}(\mathbf{v} - \mathbf{w}) - \mathbf{T}]\!] \bullet \Lambda = 0 \tag{9.52}$$

This is the standard jump balance of momentum across a singular surface [e.g. in *Slattery*, 1972; *Eringen*, 1980; *Bowen*, 1989].

The second case of particular interest is when the interface is considered massless but capable of sustaining a stress. In this instance, only the terms in the second row of equation (9.51) are retained such that:

$$- \nabla^s \bullet \mathbf{S}^{sx} - (\nabla \bullet \Lambda)\, \Lambda \bullet \mathbf{S} + [\![\rho \mathbf{v}(\mathbf{v} - \mathbf{w}) - \mathbf{T}]\!] \bullet \Lambda = 0 \tag{9.53}$$

If the interface and the two fluids on each side are constitutively treated as being Stokesian, the stress tensors for the fluids are given by equation (9.23) while the stress tensor for the interface is:

$$\mathbf{S} = \sigma \mathbf{I}^s + \mathbf{s} \tag{9.54}$$

where:

- σ is surface tension,
- $\mathbf{I}^s$ is the identity tensor in the surface, $\mathbf{I} - \Lambda\Lambda \bullet \mathbf{I}$,
- $\mathbf{I}$ is the identity tensor in space, and
- $\mathbf{s}$ is the dissipative stress in the interface.

Assume further that no mass transfer occurs across the interface (i.e. $(\mathbf{v} - \mathbf{w}) \bullet \Lambda = 0$ on both sides of the interface) such that equation (9.53) becomes:

$$- \nabla^s \sigma + \Lambda \sigma (\nabla^s \bullet \Lambda) + [\![p^f]\!] \Lambda - \nabla^s \bullet \mathbf{s}^{sx} - (\nabla \bullet \Lambda)\, \Lambda \bullet \mathbf{s} - [\![\boldsymbol{\tau}]\!] \bullet \Lambda = 0 \tag{9.55}$$

If the interface is at equilibrium, the last three terms in this equation will be zero and:

$$-\nabla^s\sigma + \boldsymbol{\Lambda}\sigma(\nabla^s \bullet \boldsymbol{\Lambda}) + [\![p^f]\!]\,\boldsymbol{\Lambda} = 0 \tag{9.56}$$

Two enlightening forms of this equation arise by taking its dot product with $\mathbf{I}^s$ and $\boldsymbol{\Lambda}$, respectively to obtain:

$$\nabla^s\sigma = 0 \tag{9.57a}$$

and:

$$\sigma(\nabla^s \bullet \boldsymbol{\Lambda}) + [\![p^f]\!] = 0 \tag{9.57b}$$

The first of these equations indicates that at equilibrium, the surface gradient of surface tension in an interface will be zero. Since $\nabla^s \bullet \boldsymbol{\Lambda}$ is the inverse of the mean curvature, equation (9.57b) states that at equilibrium, the jump in pressure of the phases on each side of the interface is equal to the surface tension divided by the mean curvature of the interface. This latter quantity is classically called the capillary pressure, p^c [e.g. in *Greenkorn*, 1983; *Miller and Neogi*, 1985]; and the defining statement:

$$p^c = \sigma(\nabla^s \bullet \boldsymbol{\Lambda}) \tag{9.58}$$

is referred to as the Young-Laplace equation [for a history of the development of this equation, see *Bikerman*, 1975].

The derivation in this subsection makes use of the integration theorems to develop a momentum equation for a surface from that in space. Then, with the assumptions and restrictions clearly identified, one can show how and in what circumstances this balance reduces to one of the standard equations for microscopic multiphase flow. This derivation may be contrasted with that of *Scriven* [1960] that uses more traditional mathematical tools.

9.5 MASS BALANCE FOR AXISYMMETRIC POROUS MEDIA FLOW

In this example, the mass balance equation for axisymmetric flow in a porous medium with negligible dynamics in the vertical is developed. A saturated medium will be assumed so that the porosity is equal to the volume of fluid in the pore space. The point of departure is the spatial mass balance for a pure fluid as given by equation (9.7). The approach chosen here is to first derive the three-dimensional macroscale equation for balance of fluid mass in a saturated porous medium using averaging theorems, and then transform the vertical direction to the megascale using an integration theorem.

The approach selected illustrates that the theorems developed in this work can be applied to differential equations which themselves are the result of some averaging procedure. In other words the averaging and integration steps can be nested. Also, note that the class of theorems (i.e. averaging or integration) can be changed during successive steps. More specifically, this application develops the differential mass balance equation at the macroscale for a porous medium using averaging theorems from the $[3, (3, 0), 0]$ family and then obtains a vertically megascopic result by applying integration theorems from the $[3, (0, 0), 1]$ family. As an alternative, one could simply apply averaging theorems from the $[3, (2, 0), 1]$ family to the microscopic mass balance equation to obtain the desired balance law via a one step process.

Equation (9.7) is the mass conservation equation for a single-phase fluid at the microscale. To transform it to the three-dimensional continuum equation for flow in a porous media requires application of $T[3, (3, 0), 0]$, equation (8.4), to the time derivative in equation (9.7) and $D[3, (3, 0), 0]$, equation (8.2), to the divergence term. Theorems from the $[3, (3, 0), 0]$ family are selected because they transform equations that are microscopic in space to equations that are macroscopic in space. Application of these theorems to equation (9.7) where the fluid is the α-phase and the porous matrix is the β-phase results in:

$$\frac{\partial}{\partial t}\int_{V_\alpha} \rho dV + \nabla \bullet \int_{V_\alpha} \rho \mathbf{v} dV + \int_{S_{\alpha\beta}} \rho(\mathbf{v} - \mathbf{w}) \bullet \mathbf{n} dS = 0 \tag{9.59}$$

If one assumes that no mass is exchanged between the α-phase fluid and the β-phase solid, the last integral is this expression is zero. Simplification of the notation in equation (9.59) using the averages defined in equations (9.12a) and (9.12b) yields:

$$\frac{\partial(\langle\rho\rangle V_\alpha)}{\partial t} + \nabla \bullet (\langle\rho\rangle \bar{\mathbf{v}} V_\alpha) = 0 \tag{9.60}$$

When the matrix is saturated, the porosity is equal to the volume fraction of the REV occupied by the α-phase such that:

$$\epsilon = \frac{V_\alpha}{V} \tag{9.61}$$

where:

ϵ is the porosity,

V_α is the portion of the REV occupied by the α-phase, and

V is the volume of the REV.

For the $[3,\ (3,0),0]$ family of theorems, V is taken to be constant in space and time. Therefore division of equation (9.60) by V gives:

$$\frac{\partial(\langle\rho\rangle\epsilon)}{\partial t}+\nabla\bullet(\langle\rho\rangle\bar{\mathbf{v}}\epsilon)=0 \qquad \textbf{(9.62)}$$

Equation (9.62) is the spatial mass balance for the fluid in a saturated porous medium with no internal sources or sinks of mass and no phase change.

The next step in the derivation to obtain the axisymmetric flow equation is to integrate equation (9.62) over the vertical coordinate in the porous medium formation, or aquifer, using an integration theorem. Since equation (9.62) is a spatial balance law, albeit a fully macroscopic one, and integration will be over one coordinate direction, the required integration theorems will be in the $[3,\ (0,0),1]$ family. In this case integration is over a straight line and thus the theorem for integration of the time derivative is given by $T[3,\ (0,0),1]$, equation (7.40), while the divergence term is integrated using $D[3,\ (0,0),1]$, equation (7.38) (See figure 7.6 for a diagram and notation). Application of these theorems to equation (9.62) yields:

$$\frac{\partial}{\partial t}\int_L\langle\rho\rangle\epsilon dL+\nabla^s\bullet\int_L\langle\rho\rangle\epsilon\bar{\mathbf{v}}^s dL+\int_L\langle\rho\rangle\epsilon(\nabla^s\bullet\mathbf{\Lambda})\,\mathbf{\Lambda}\bullet\bar{\mathbf{v}}^c dL$$

$$+\left(\frac{\langle\rho\rangle\epsilon(\bar{\mathbf{v}}-\mathbf{w})\bullet\mathbf{n}^*}{\mathbf{e}\bullet\mathbf{n}^*}\right)\Bigg|_{\text{ends}}=0 \qquad \textbf{(9.63)}$$

The third integral in this equation will be neglected here because both the vertical velocity, $\bar{\mathbf{v}}^c$, and the change of direction of $\mathbf{\Lambda}$ with areal position are considered small. If the vertical variations of density, velocity, and porosity are negligible (or else if these quantities are expressed in terms of their vertical averages), equation (9.63) reduces to:

$$\frac{\partial(\langle\rho\rangle\epsilon L)}{\partial t}+\nabla^s\bullet(\langle\rho\rangle\epsilon\bar{\mathbf{v}}^s L)+\left(\frac{\langle\rho\rangle\epsilon(\bar{\mathbf{v}}-\mathbf{w})\bullet\mathbf{n}^*}{\mathbf{e}\bullet\mathbf{n}^*}\right)\Bigg|_{\text{ends}}=0 \qquad \textbf{(9.64)}$$

where L is the thickness of the porous medium and may be a function of both space and time. In this equation, the last term accounts for leakage into and out of the aquifer at its top and bottom boundaries.

For the case where the flow in an aquifer is radially symmetric, such as for flow to a well in a homogeneous system, the surface divergence operator in equation (9.64) may be reduced to a a more convenient form. From equation (2.50c) and the geometric information provided in Table 2.2, surface divergence in $r-\theta$ cylindrical coordinates may be obtained such that:

$$\nabla^s \bullet (\langle\rho\rangle\epsilon\bar{\mathbf{v}}^s L) = \frac{\partial(\langle\rho\rangle\epsilon\bar{v}_r L)}{\partial r} + \frac{1}{r}\frac{\partial(\langle\rho\rangle\epsilon\bar{v}_\theta L)}{\partial\theta} + \frac{\langle\rho\rangle\epsilon\bar{v}_r L}{r} \quad (9.65)$$

where $\bar{v}_r = \bar{\mathbf{v}}^s \bullet \mathbf{e}_r$ and $\bar{v}_\theta = \bar{\mathbf{v}}^s \bullet \mathbf{e}_\theta$. For the radially symmetric case, variation in the θ-direction is negligible so that the second term on the right side of equation (9.65) is zero and the remaining terms may be rearranged to obtain:

$$\nabla^s \bullet (\langle\rho\rangle\epsilon\bar{\mathbf{v}}^s L) = \frac{1}{r}\frac{\partial(r\langle\rho\rangle\epsilon\bar{v}_r L)}{\partial r} \quad (9.66)$$

Thus equation (9.64) takes the form:

$$\frac{\partial(\langle\rho\rangle\epsilon L)}{\partial t} + \frac{1}{r}\frac{\partial(r\langle\rho\rangle\epsilon\bar{v}_r L)}{\partial r} + \left(\frac{\langle\rho\rangle\epsilon(\bar{\mathbf{v}} - \mathbf{w}) \bullet \mathbf{n}^*}{\mathbf{e} \bullet \mathbf{n}^*}\right)\Bigg|_{\text{ends}} = 0 \quad (9.67)$$

The first term accounts for the accumulation of mass in the system, the second represents the net advective flux in the radial direction, and the last term is the flux of mass across the system boundaries at the top and bottom of the aquifer.

9.6 DERIVATION OF ADDITIONAL THEOREMS

Although the 128 integration and averaging theorems provided in Chapters 7 and 8 provide of a rather extensive set, the possibility exists that other theorems may be needed for particular applications because of expressions that arise in analyzing a problem. The needed theorems may be obtained directly using the derivation procedures involving the generalized functions, or they may be obtained as particular cases or combinations of some of the theorems provided. In this section, the latter approach will be applied, for purposes of illustration, to obtain two rather well-known formulas.

9.6.1 Stokes' Theorem

Stokes theorem relates the integral over a surface, S, of the normal component of the curl of a vector to the integral of the tangential component of the vector along the curve, C, bounding the surface. One way to obtain this theorem is to make use of equation (7.14), the divergence theorem $D[3, (0,0), 2]$, for the case when the vector $\mathbf{f}$ is set equal to the cross product $\mathbf{a} \times \mathbf{n}$. By this definition, the vector component of $\mathbf{f}$ in the $\mathbf{n}$-direction, indicated as $\mathbf{f}^c = \mathbf{nn} \bullet (\mathbf{a} \times \mathbf{n})$, will be zero. Thus equation (7.14) becomes:

$$\int_S \nabla \bullet (\mathbf{a} \times \mathbf{n})\, dS = -\int_S (\mathbf{n} \bullet \nabla^c \mathbf{n}) \bullet (\mathbf{a} \times \mathbf{n})\, dS + \int_C \boldsymbol{\nu}^* \bullet (\mathbf{a} \times \mathbf{n})\, dC \quad (9.68)$$

Standard mathematical identities involving dot and cross products [e.g. *Zill and Cullen*, 1992] show that:

$$\nabla \bullet (\mathbf{a} \times \mathbf{n}) = \mathbf{n} \bullet (\nabla \times \mathbf{a}) - \mathbf{a} \bullet (\nabla \times \mathbf{n}) \quad \textbf{(9.69a)}$$

$$(\mathbf{n} \bullet \nabla^c \mathbf{n}) \bullet (\mathbf{a} \times \mathbf{n}) = -\mathbf{a} \bullet (\mathbf{n} \bullet \nabla^c \mathbf{n} \times \mathbf{n}) \quad \textbf{(9.69b)}$$

$$\boldsymbol{\nu}^* \bullet (\mathbf{a} \times \mathbf{n}) = \mathbf{a} \bullet (\mathbf{n} \times \boldsymbol{\nu}^*) \quad \textbf{(9.69c)}$$

Substitution of these three relations into equation (9.68) and regrouping of terms yields:

$$\int_S \mathbf{n} \bullet (\nabla \times \mathbf{a}) \, dS = \int_S \mathbf{a} \bullet (\nabla \times \mathbf{n} + \mathbf{n} \bullet \nabla^c \mathbf{n} \times \mathbf{n}) \, dS + \int_C \mathbf{a} \bullet (\mathbf{n} \times \boldsymbol{\nu}^*) \, dC \quad \textbf{(9.70)}$$

However, equation (2.15b) indicates that the quantity in parentheses in the first integral on the right side is equal to zero. Additionally, the cross product of the orthogonal vectors **n** and $\boldsymbol{\nu}^*$ is equal to $\boldsymbol{\lambda}$, the tangent to C. Thus equation (9.70) reduces to:

Stokes' Theorem

$$\int_S \mathbf{n} \bullet (\nabla \times \mathbf{a}) \, dS = \int_C \mathbf{a} \bullet \boldsymbol{\lambda} \, dC \quad \textbf{(9.71)}$$

9.6.2 Alternative Transport Theorems for a Surface

Suppose a transport theorem is desired for a surface, S, but no integration over the boundary of the surface, C, is to appear. This theorem may be obtained by adding divergence theorem $D[3, (0, 0), 2]$, equation (7.14),with $\mathbf{f} = \mathbf{w}f$ to transport theorem $T[3, (0, 0), 2]$, equation (7.16), such that the integrals over C cancel:

$$\int_S \left.\frac{\partial f}{\partial t}\right|_{\mathbf{x}} dS + \int_S \nabla \bullet (\mathbf{w}f) \, dS = \frac{d}{dt}\int_S f dS + \int_S \nabla^c \bullet (\mathbf{w}^c f) \, dS$$

$$- \int_S \mathbf{n} \bullet \nabla^c \mathbf{n} \bullet \mathbf{w}^s f dS - \int_S \mathbf{w}^c \bullet \nabla^c f dS \quad \textbf{(9.72)}$$

The divergence in the second integral on the right side of this equation may be expanded to show that:

$$\nabla^c \bullet (\mathbf{w}^c f) = (\mathbf{nn} \bullet \nabla) \bullet (\mathbf{nn} \bullet \mathbf{w}f) = \mathbf{n} \bullet \nabla \mathbf{w} \bullet \mathbf{n} f + \mathbf{n} \bullet \nabla^c \mathbf{n} \bullet \mathbf{w}^s f + \mathbf{w}^c \bullet \nabla^c f$$

$$\textbf{(9.73)}$$

Substitution of this expansion into equation (9.72) allows for cancellation of some terms. Rearrangement of the surviving terms then gives:

$$\frac{d}{dt}\int_S f dS = \int_S \left.\frac{\partial f}{\partial t}\right|_{\mathbf{x}} dS + \int_S \nabla\cdot(\mathbf{w}f)\, dS - \int_S \mathbf{n}\cdot\nabla\mathbf{w}\cdot\mathbf{n} f dS \qquad \textbf{(9.74)}$$

However, $\nabla^s\cdot\mathbf{w} = \nabla\cdot\mathbf{w} - \mathbf{n}\cdot\nabla\mathbf{w}\cdot\mathbf{n}$ so that equation (9.74) may alternatively be written as:

First Alternative Transport Theorem for a Surface

$$\frac{d}{dt}\int_S f dS = \int_S \frac{df}{dt} dS + \int_S (\nabla^s\cdot\mathbf{w}) f dS \qquad \textbf{(9.75a)}$$

where the definition of the total derivative has been employed such that:

$$\frac{df}{dt} = \left.\frac{\partial f}{\partial t}\right|_{\mathbf{x}} + \mathbf{w}\cdot\nabla f \qquad \textbf{(9.75b)}$$

An interesting variation of this theorem pertains for the case where the scalar f is defined as the dot product of a vector $\mathbf{a}$ with the unit vector $\mathbf{n}$ that is normal to the surface S. With $f = \mathbf{a}\cdot\mathbf{n}$, equation (9.75a) becomes:

$$\frac{d}{dt}\int_S \mathbf{a}\cdot\mathbf{n} dS = \int_S \frac{d(\mathbf{a}\cdot\mathbf{n})}{dt} dS + \int_S (\nabla^s\cdot\mathbf{w})\, \mathbf{a}\cdot\mathbf{n} dS \qquad \textbf{(9.76)}$$

Application of the chain rule to the first integral on the right side and substitution of equation (6.49), $d\mathbf{n}/dt = -(\nabla^s\mathbf{w})\cdot\mathbf{n}$, to eliminate $d\mathbf{n}/dt$ yields:

Second Alternative Transport Theorem for a Surface

$$\frac{d}{dt}\int_S \mathbf{a}\cdot\mathbf{n} dS = \int_S \frac{d\mathbf{a}}{dt}\cdot\mathbf{n} dS - \int_S \mathbf{a}\cdot(\nabla^s\mathbf{w})\cdot\mathbf{n} dS + \int_S (\nabla^s\cdot\mathbf{w})\, \mathbf{a}\cdot\mathbf{n} dS \qquad \textbf{(9.77)}$$

This theorem has been presented, with its proof left as an exercise, in *Aris* [1962].

9.7 CONCLUSION

The objective of this chapter has been to demonstrate the use of the integration and averaging theorems found in Chapters 7 and 8 for changing the point representation of physical problems to a representation at some larger scale. Advantages of this approach over the more traditional control volume approach are:

the procedure starts at the microscale and systematically proceeds to the macroscale/megascale so that terms representing physical processes at the microscale are carried along to the larger scale; effects of curvature in the system geometry are included naturally; insight into the physics of the problem at the larger scale is not buried in constitutive relations invoked at the larger scale. While the procedure is straightforward, some care is required in choosing the appropriate theorems and in properly interpreting the terms that result from the integration process. Also, potential simplification of the general form of the mathematically generated balance law requires a clear understanding of the physics of the process under consideration. Examples in this chapter have highlighted some of the most important techniques. Other applications (e.g. averaging of the energy and entropy balance equation for a multiphase system) for three-dimensional porous media systems can be found in the papers by *Hassanizadeh and Gray* [1979a, 1979b]. An application to surfaces in a multiphase system can be found in *Gray and Hassanizadeh* [1989] while *Stone* [1990] has developed the equation for species transport along a deforming surface.

One important consideration that arises is the choice of the integration theorems from Chapter 7 or the averaging theorems from Chapter 8. Selection of the appropriate theorem depends on the scales of the processes under consideration and the degree of filtering, or averaging, that is desired. In some cases, especially for single phase systems, as was shown in the first example in Section 9.2, the result obtained by either integration or averaging theorems is similar although the resulting terms must be interpreted differently. In other applications, the flexibility to select between an integration or an averaging theorem may not be available. The following general guidelines for theorem selection should be considered.

In practice, all of the averaging theorems in Chapter 8 act as a filter to smooth a noisy function, i.e. the function is integrated over a region to produce an average value multiplied by the extent of the region. The macroscale value of the function is taken to be the averaged value applied at the centroid of the REV. As the REV is located throughout the domain of interest, a continuous representation of the macroscale, or filtered, function is obtained. It is important to note that for the averaging theorems developed here, averaging is performed over a portion of the space within an REV, over the surface contained within the REV, or over a curve contained within the REV. For the theorems of Chapter 8, REV's are selected to be constant radius spheres (for full macroscopization), constant radius cylinders (for macroscopization in two dimensions and megascopization in one dimension), or constant thickness slabs (for macroscopization in one dimension and megascopization in two dimensions). These geometries are selected for convenience, rather than out of need, with the constraints that 1) macroscopic length scales must be constant in space and time for a particular application; and 2) when partial megascopic averaging is employed, the radius of curvature of the change in orientation of the REV must be much larger than the macroscale. The possibility exists for theorems to be developed that allow for less restrictively defined averaging regions [*Gray*, 1983], but the utility of averaged equations that retain macroscopic dependence on the shape and orientation of the averaging volume seems to be limited. If one is interested in obtaining continuum equations for a multiphase system where the non-overlapping microscopic phases, interfaces, and contact lines are to be modeled as overlapping continua at the macroscale, the theorems of Chapter 8 must be used.

In contrast, the integration theorems in Chapter 7 apply only to a continuum. Separate phases, indeed discontinuities in a function, are not allowed within the region of integration while applying these theorems. Additionally, the theorems in Chapter 7 are more precise in that they apply when the functions being considered have continuous first derivatives. The theorems of Chapter 8 are, in fact, approximations that hold subject to constraints on the relative sizes of the length scales (microscale must be much less than the macroscale which in turn must be much less than the megascale). and the properties of the averaging volumes mentioned above.

In summary, the application of generalized functions as presented here to obtain averaging and integration theorems is a powerful tool for development of integration and averaging theorems. These functions have been used to obtain an extensive set of 128 integration and averaging theorems. Although this set of theorems is adequate for many applications involving dynamics of single and multiphase systems in space, on arbitrary surfaces, and along arbitrary curves, applications may arise where variations on these theorems are required. These variations may be developed either from basic principles by applying the generalized functions and making use of the identities of Chapter 6 or by combining theorems from a single family and making use of some standard relations from calculus. In either event, the theorems developed allow problems based on the basic laws of physics to be analyzed systematically and with rigor.

9.8 REFERENCES

Abbot, M. B., *Computational Hydraulics*, Pitman, London, 1979.

Aris, R., *Vectors, Tensors, and the Basic Equations of Fluid Mechanics*, Prentice-Hall, Englewood Cliffs, 1962.

Bikerman, J. J., Theories of capillary attraction, *Centaurus*, **19**(3), 182-206, 1975.

Bird, R. B., W. E. Stewart, and E. N. Lightfoot, *Transport Phenomena*, John Wiley and Sons, New York, 1960.

Cunge, J. A., F. M. Holly, Jr., and A. Verwey, *Practical Aspects of Computational River Hydraulics*, Pitman, London, 1980.

Fetter, C. W., *Applied Hydrogeology*, Second Edition, Merrill Publishing Company, Columbus, Ohio, 1988.

Gray, W. G., Local volume averaging of multiphase systems using a non-constant averaging volume, *Intl. J. Multiphase Flow*, **9**(6), 755-761, 1983.

Gray, W. G. and S. M. Hassanizadeh, Averaging theorems and averaged equations for transport of interface properties in multiphase systems, *Intl. J. Multiphase Flow,* **15**(1), 81-95, 1989.

Hassanizadeh, S. M., and W. G. Gray, General conservation equations for multiphase systems: 1. averaging procedure, *Adv. Water Resources*, **2**, 131-144, 1979a.

Hassanizadeh, S. M., and W. G. Gray, General conservation equations for multiphase systems: 2. mass, momentum, energy, and entropy," *Adv. Water Resources*, **2**, 191-203, 1979b.

Jansen, P. Ph., L. van Bendegom, J. van den Berg, M. de Vries, and A. Zanen, *Principles of River Engineering, The Non-tidal Alluvial River*, Pitman, London, 1979.

John, J. E. A., and W. L. Haberman, *Introduction to Fluid Mechanics*, Second Edition, Prentice Hall, Englewood Cliffs, 1980.

Le Méhauté, B., *An Introduction to Hydrodynamics and Water Waves*, Springer-Verlag, New York, 1976.

Malvern, L. E., *Introduction to the Mechanics of a Continuous Medium*, Prentice-Hall, Englewood Cliffs, 1969.

Stone, H. A., A simple derivation of the time-dependent convective-diffusion equation for surfactant transport along a deforming interface, *Physics of Fluids A*, **2**(1), 111-112, 1990.

Whitaker, S., *Introduction to Fluid Mechanics*, Prentice-Hall, Englewood Cliffs, 1968.

White, F. M., *Fluid Mechanics*, McGraw-Hill, New York, 1979.

Zill, D. G., and M. R. Cullen, *Advanced Engineering Mathematics*, PWS-Kent, 1992.

INDEX